The Politics of Invisibility

The Politics of Invisibility

Public Knowledge about Radiation Health Effects after Chernobyl

Olga Kuchinskaya

The MIT Press
Cambridge, Massachusetts
London, England

This book was set in Stone Sans and Stone Serif by the MIT Press.

Library of Congress Cataloging-in-Publication Data

Kuchinskaya, Olga, 1975–
The politics of invisibility : public knowledge about radiation health effects after Chernobyl / Olga Kuchinskaya.
  pages cm. — (Infrastructures)
Includes bibliographical references and index.
ISBN 978-0-262-02769-4 (hardcover : alk. paper)
ISBN 978-0-262-54886-1 (paperback)
1. Chernobyl Nuclear Accident, Chernobyl, Ukraine, 1986—Health aspects. 2. Chernobyl Nuclear Accident, Chernobyl, Ukraine, 1986—Social aspects. 3. Communication in medicine—Belarus. 4. Communication in medicine—Europe, Eastern. 5. Health risk assessment—Government policy—Belarus. 6. Health risk assessment—Government policy—Europe, Eastern. 7. Radiation victims—Belarus—Attitudes. 8. Radiation victims—Europe, Eastern—Attitudes. 9. Health surveys—Belarus. 10. Health surveys—Europe, Eastern. I. Title.
RA569.K88   2014
363.17'99094777—dc23
2013048602

Contents

Preface

This research started in 2003, when I discovered online reports and press releases by the United Nations Scientific Committee on the Effects of Atomic Radiation (UNSCEAR) claiming, in essence, that Chernobyl was a myth.[1] In its objective scientific voice, UNSCEAR argued that there was no evidence that radioactive fallout from the 1986 Chernobyl nuclear accident had significant effects on the health of the affected populations. Only one disease was linked to radiation exposure: thyroid cancer in children. All other increases in health problems were blamed on radiophobia (fear of radiation), stress resulting from the collapse of the Soviet Union, and the degradation of people's living conditions at the time. I had taken it as indisputable that Chernobyl had devastating consequences and that Belarus, the country in which I grew up, was most affected by it. The UNSCEAR reports confronted me with the fact that what I considered *obvious* from my perspective was interpreted as *nonexistent* from a different—expert and institutionally powerful—position; their judgment was buttressed by claims to objectivity.

I called my family and friends in Belarus and asked them if it was true that Chernobyl had no effects. My family and friends had relatively limited exposure to the consequences of Chernobyl. They resided in Minsk, more than a hundred miles from most of the heavily affected areas, but they were closer to it than I was, living in the United States. They assured me that "everybody knows" there were 16,000, 30,000, or even 100,000 victims. According to them, the accident also resulted in a great rise in health problems in Belarus, and this rise could be observed in numbers. Those numbers were "out there," they said, and told me to "look them up."

Some basic online and database searches turned up more press releases and reports from conferences sponsored by international organizations, including the International Atomic Energy Agency (IAEA). But my searches also revealed articles by Belarusian and international scientists with claims about various specific health effects related to the radiation released from Chernobyl, including a range of noncancer, somatic problems. My attempt to delve into the data and compare the evidence was not satisfactory; the data and findings were fragmentary, and much seemed to be missing. Connections between UNSCEAR and the international nuclear industry became apparent rather quickly; it was not surprising that nuclear industry experts might be motivated to downplay the perceived consequences of a nuclear accident. But I still did not have an answer to my original question about the *actual* scope and nature of the health effects.

Trying to learn more from Chernobyl reports was making me increasingly uncertain. I even thought that perhaps uncertainty was part of the state of knowledge on this issue, except that few seemed to share that uncertainty. The reports appeared certain, even if conflicting or difficult to agree with. The people I spoke to in Belarus remained perfectly certain, though admittedly not very knowledgeable about the details of scientific research on Chernobyl.

My search for numbers led to another question. I wanted to know what the consequences were from the perspective of those most affected by the accident. But I also became puzzled by how those affected learn about the contamination and its health effects. Radiation cannot be perceived by unaided human senses. It is invisible. Unless the doses are extremely high, it causes no immediate bodily problems. The health effects are often delayed in time and are not radiation-specific.[2] A person is not marked by radiation exposure in the same way as when, for example, he or she falls off a bicycle and bruises a leg. The effects of radiation, along with the causal connections between exposure and resulting health problems, are not immediately observable.

The following thought experiment illustrates how the imperceptibility of environmental hazards such as radiation affects how we learn about them. Imagine that the room you are currently in, or the lunch in front of you, has a source of increased radioactivity. You cannot see it, taste it, smell it, or hear it, and it doesn't hurt. You might be filled with a vague sense of dread, but without the input of some experts with their theories and equipment, you have no way of knowing whether you are really in danger. Without the experts and their equipment, we are left with impressions and no direct experience of the reality to verify them. Now imagine that some of the people who frequent this room develop health problems several years later. Not only can we not notice increased levels of radiation on our own, we also cannot easily establish their causal relationship to later health issues unless someone (an expert, the media, or a person in the know) raises this possibility.

This lack of immediate experiential confirmation of either radiation or its effects makes it more difficult for the Chernobyl-affected populations to notice the danger of radiation. But questioning how laypeople learn about radiation and its effects—or about any number of invisible dangers around us—might be just the first step. We should also consider whether and how this "lay invisibility" affects the practices of professionals (scientists, reporters, and government officials) as they try to make sense of this and other invisible threats and represent them in public discussions. The heart of the matter is how we discuss environmental hazards like radiological contamination in the public sphere when even the affected groups see little experiential evidence for the problem. I thus frame the main question for my research broadly: How is the public discussion of Chernobyl shaped? Or, more specifically, how have the interactions among different perspectives—of various groups of laypeople, experts, and policy makers—been shaping what we interpret as the consequences of Chernobyl? The issue of the lay imperceptibility of hazards also raises the question of how we know that scientific research about these hazards, journalistic reporting, or even government-prescribed protection measures are adequate. We often have faith that "the truth will prevail," but what social mechanisms guarantee that, when even the affected populations neither see the contamination nor necessarily recognize its effects on their health?

I repeatedly heard the sentiment that "time will tell" from Belarusian scientists, whose theories and data were being ignored or discredited by the UN reports. Research in social studies of science tells us that there is no strict correspondence between the seriousness of a hazard and the amount of attention it receives publicly. Great effort can be put into *not* constructing risks.[3] A priori, nothing precludes the effects of a major nuclear accident from being ultimately obscured and ignored. Therefore, instead of trying to find out facts about Chernobyl's consequences, I have sought to examine the issues related to the creation of knowledge about it. My goal is to consider what social mechanisms ensure that we learn the full story.

Acknowledgments

I am deeply grateful to everybody in Belarus who participated in this research by sharing his or her perspective and by helping establish further connections. I especially thank the following individuals, who were particularly generous with their time and information year after year: the late Vassily Nesterenko, Vladimir Babenko, and Mikhail Malko. I also received support from many other individuals who worked or continue to work in Chernobyl-related research institutes, nongovernmental organizations and international projects, and government organizations, including Comchernobyl and Minzdrav. Following the conventions for social research, most names in this book have been changed or omitted, except when the individuals are easy to identify because of their professional activities or publications.

I thank my mentors at the University of California, San Diego, and my colleagues from the University of Pittsburgh and elsewhere who discussed this project with me, provided useful advice, and offered comments on earlier versions of this text: Daniel Hallin, Susan Leigh Star, Geoffrey Bowker, Andrew Lakoff, Steven Epstein, Valerie Hartouni, Vincent Rafael, Ravi Rajan, Mary Gray, Wenda Bauchspies, Kelli Moore, Kelly Gates, Astrid Sahn (who also shared her personal research materials), Andrei Stepanov, Gary Fields, Mary Judson, Jericho Burg, Lucy Tatman, Tom Kane, Gordon Mitchell, John Lyne, Ronald Zboray, Mary Zboray, Gabriella Lukács, Paul Josephson, Soraya Boudia, Jennifer Bails, and Olga Maslovskaya. Many thanks to Gabrielle Hecht for her perceptive feedback on the full draft of this manuscript. None of the individuals mentioned above bear any responsibility for the analysis that follows; their advice helped make this text better. I am particularly grateful to Geof Bowker and the late Susan Leigh Star, whose own research inspired much of this project.

Many thanks to the MIT Press, and especially to Marguerite Avery, Katie Persons, and Deborah Cantor-Adams for their guidance and patience. I thank Audra Wolfe and Judith Antonelli for their editorial assistance.

This research would not have happened without help from my family and friends, who provided active support throughout parts of this project and showed much patience at other stages. I thank Olga Maslovskaya, Nataliya Lvovich, Jericho Burg, Lucy Tatman, Irina Skoblya, Nina Zelutkina, and Tatiana Haplichnik; my mother-in-law, Natallia Rybianets; my brother, Arseni Kuchinsky; my parents, Tatiana Kuchinskaya and Gennadi Kuchinsky; and my son, Sasha. I am deeply grateful to my husband, Ivan Graman, for his unwavering support and for so much more.

Two chapters appeared in earlier versions as the articles "Articulating the Signs of Danger: Lay Experiences of Post-Chernobyl Radiation Risks and Effects," in *Public Understanding of Science* 20, no. 3 (2011): 405–21, and "Twice Invisible: Formal Representations of Radiation Danger," in *Social Studies of Science* 43, no. 1 (2013): 78–96.

Introduction

The 1986 Chernobyl nuclear accident is one in a list of many: Sellafield (England, 1957), Three Mile Island (Pennsylvania, 1979), and Fukushima Daiichi (Japan, 2011), along with numerous minor accidents. As of this writing, Chernobyl is the largest accident in the list. The fallout from the accident covered vast areas in the Northern Hemisphere, especially in Europe. Belarus, which at the time was a Soviet republic north of Ukraine and its Chernobyl nuclear power plant, received most of the fallout (Ukraine itself and areas of the Russian Federation were also heavily affected). According to the official numbers, 23 percent of the Belarusian territory was covered with long-lasting radioactive isotopes.[1]

The 2005 commemorative calendar published by the Belarusian government and the United Nations Development Programme (UNDP) tells us that 2.3 million people were affected in Belarus; 135,000 residents and 415 communities were permanently relocated.[2] Plants, factories, farms, schools, and medical facilities had to be closed in the most contaminated areas. The economic damage from the Chernobyl disaster amounted to $235 billion, "which amounts to 32 annual [Belarusian] budgets for 1986, the year when the accident occurred," the calendar stated. This description resembles descriptions of natural disasters, with their evident and often spectacularly dramatic consequences. Yet radiological fallout does not destroy houses. There is no immediately visible destruction. Contaminated forests and communities look exactly like uncontaminated ones. For the vast majority of the affected, no health problems are immediately obvious.

The official numbers describing the accident reveal some consequences of Chernobyl, but these numbers do not tell us how anybody came to identify these consequences. These numbers do not tell us about the scientific and administrative assessments and decision making that had to follow the fallout, as well as many effects of the decisions that were made—or not made. The broad purpose of this book is to critically examine how we recognize and learn to respond to imperceptible environmental hazards, when neither the hazard nor its consequences are immediately observable. The analysis in this book is based on the following insight: the imperceptibility of Chernobyl radiation by the human senses means that individuals' experience of it is always highly mediated.[3] There is no direct component to one's experience of radiation, no simple sensory (and commonsense) evidence of the kind referred to in the old joke, "How do you know that something exists? Kick it really hard." Our experience of imperceptible hazards is always necessarily mediated by measuring equipment, maps, and other ways to visualize it, but also with narratives. Different ways of representing Chernobyl can make radiation and its effects observable and publicly visible, or they can make them unobservable and publicly nonexistent.

How these representations are produced matters. Much of the analysis offered in this book describes the *production of invisibility* of Chernobyl's consequences—that is, the practices of producing representations that limit public visibility of Chernobyl radiation and its health effects.[4] This is a question of what does not appear in public discussions. Simply put, I describe the double twist: how imperceptible hazards, such as radiation, are made publicly invisible.

Limiting public visibility of Chernobyl radiation prevents the construction of links between radiation and its health effects, which can in turn be described as the "social construction of ignorance."[5] As a result, what could under other circumstances be identified as Chernobyl radiation health effects dissolves into individual health problems of unspecific origins. The production of public invisibility of imperceptible hazards can be thought of as a process opposite to that of the discovery of microbes. Microbes also escape our perception, and they existed before Louis Pasteur's experiments, but not as socially recognized actors that can cause somebody to fall sick and that must be dealt with through a variety of precautionary measures. Microbes had to be *made visible*.[6] I demonstrate that making something visible is not a one-way process: imperceptible phenomena can also be made publicly invisible, and this production of invisibility is both work and a consequence of particular structural conditions.

I approach post-Chernobyl radiological contamination as one example of modern environmental risks. These risks pose particular challenges; German sociologist Ulrich Beck gives us a good starting point for considering them. Radiation is one of Beck's paradigmatic examples, although one can find many similar imperceptible risks in our daily lives, including pesticides in our food, chemical toxins in our environment, and even global warming.

The problem of modern risks for Beck is at least partly a question of the role of experts and the future of democratic decision making. Identifying these hazards as such requires establishing complex causal connections. To use a common expression, we need to "connect the dots." Science experts play a disproportionately significant role in that process, from establishing the fact of exposure to evaluating health effects. According to Beck, risks not recognized scientifically "do not exist legally, medically, technologically, or socially, and they are thus not prevented, treated or compensated for."[7] Beck also argues that laypeople, even when directly exposed to risks, are "culturally blind" to them. Perhaps if radioactivity made us itch, one could make sense of it more easily—it would fall "within the orbit of cultural experience." Without such natural perceptibility or some "culturally manufactured perceptibility," simply judging for ourselves, without replying on experts, has become impossible.[8]

Beck's discussion of Chernobyl gives us more reasons for concern. He ponders the consequences of Chernobyl in the context of Germany and argues that "the number of Chernobyl dead will never be counted." For a person wishing to learn about the consequences of the accident, the experience might bring to mind Franz Kafka's story "The Trial." The labyrinth of bureaucratic, organizational complexity and irresponsibility makes such attempts self-defeating. At the end of Beck's ironic exposé, the person acknowledges his own naive mistake of bringing up such a difficult and complex matter in the first place.[9] I am reminded of this exposé whenever some students in my environmental communication class decide to learn more about, perhaps even get to the bottom of, a particular environmental issue, such as what exactly is in the air they breathe in Pittsburgh or how fracking affects the water in their home communities. They often report feeling overwhelmed and discouraged by complexity—that of the phenomenon itself, but also scientific and bureaucratic complexity. I certainly had that experience as I sought to learn more about the consequences of Chernobyl.

This book does not offer a definitive account of Chernobyl's health effects. Instead, I look at the nature of the challenges and the possible systematic problems of "seeing" and learning about imperceptible environmental risks. I trace systematic omissions and biases, along with their effects on identifying these risks. Neither public visibility nor invisibility of environmental hazards—even those as massive as post-Chernobyl fallout—is a natural, spontaneous process. Recognizing radiation danger and its effects is not necessarily a linear process, leading to increasingly more accurate and more comprehensive understanding. Similarly, risks disappear in public discussion not necessarily because the danger has abated. Public recognition of hazards can be altered, redefined, and reframed, or it can shift with changes of historical context. Indeed, as we will see, the scope of recognized post-Chernobyl risks has fluctuated historically.

The commemorative calendar mentioned above says nothing about the scope of radiation-related health effects. Such an omission is not unusual for the official descriptions of the consequences of Chernobyl in Belarus. Yet ignoring the health effects of radiation exposure—and potentially ignoring them systematically—would make it an issue of social justice. Somebody's past, present, or future suffering might be ignored. From this perspective, the invisibility of imperceptible hazards connotes not disregard for reality as much as injustice and a lack of social mechanisms to ensure that representations of imperceptible risks are adequate—that is, socially just.

Let us now consider what factors complicate or render the public visibility of imperceptible risks less desirable.

Catastrophe after the Accident

The late Vassily Nesterenko, one of the most outspoken nuclear experts in Belarus, told me, in a personal interview in 2005, "The problem is the nature of the problem. No government would be able to take adequate measures if faced with a situation like this, where the effort required far exceeds the state's capacity." The problem of Chernobyl for Belarus—what happened after the explosions at the Chernobyl nuclear power plant—can be described as a chronic postaccident catastrophe.

Radiological contamination was the direct consequence of the accident, but it was not the only consequence. Indeed, what was officially emphasized as Chernobyl's consequences in Belarus had more to do with the outcomes of the early mitigation of the accident, including the evacuation and resettlement of hundreds of thousands of people and the resulting destruction of the local infrastructures.[10] The Chernobyl accident was officially given the status of *catastrophe* in Belarus in 1989, after three years of isolated measures and attempts to obscure the extent of the accident by the Soviet authorities. During 1989–1991, Chernobyl was at its most visible in the Belarusian media and official discourse, and in Belarus it was the discourse of tragedy. The direct consequences of the accident could not be fully fixed, and attempting to fix them—including through evacuation, resettlement, and monetary compensation to keep people from abandoning the affected areas—produced consecutive ripples of other consequences, adding to and transforming the original problem of chronic and pervasive radiological contamination.

Put another way, the problem is that recognizing and mitigating the radiological aftermath of the accident created massive waves of secondary effects. In an interview in the newspaper *Gomel'skaya Pravda* on August 1, 1990, Mikhail Savitsky, a prominent Belarusian artist, explained the interpretation of Chernobyl as a catastrophe by referring to consequences that essentially could not be fixed:

It appears to me that the word "accident" does not fit what happened in Chernobyl. Based on its consequences, it's a catastrophe; and in relationship to people's fates, it's a tragedy. An accident can be taken care of [fixed], "liquidated." But catastrophe ... if it occurs, what has happened is not fixable. Here, all the paths that we've been traveling have intersected with each other, collided together.

Responses to the predicaments created by the accident changed during different historical periods. Indeed, the history of articulating and attempting to mitigate the effects of Chernobyl is the history of political transformation in the region, where significant changes to (in)visibility correspond to historical shifts in politics and the economy. The Soviet Union's collapse in 1991 made the socioeconomic circumstances worse and triggered new processes and practices in the production of Chernobyl (in)visibility. In the mid-1990s the discourse about the consequences of Chernobyl was transformed once again, facilitating the subsequent "disappearance" of the actual and potential health effects of the accident.

At the same time, the Belarusian government's approach to Chernobyl's effects was not entirely a local phenomenon. The newly independent Belarusian state lacked the resources to fulfill its program of mitigating Chernobyl's consequences; this program had been adopted during the last years of the Soviet Union in the hope of attracting international assistance (see chapter 3). Such assistance never materialized to an extent remotely close to what was needed, but the hope for help from outside affected the relationship of the state to non-Belarusian experts—especially those affiliated with the International Atomic Energy Agency (IAEA), World Health Organization (WHO), and other UN organizations—and its relationship to local experts, to the directions of local research and the support for it. As chapters 5 and 6 show, this three-way relationship of the Belarusian government, Belarusian scientists, and international organizations has changed over time, but the economic needs of the Belarusian government have ultimately made it more receptive to the perspective advocated by the IAEA and UNSCEAR. The history of articulating the effects of Chernobyl is the political history of the region, but it has not been walled off from the power influences from outside.

It might be tempting to explain the production of the invisibility of Chernobyl's consequences in Belarus simply as an inevitable result of the lack of resources—the poverty of the Belarusian state and of its people. After all, material desperation might lead one to ignore long-term problems, such as radiological contamination and its delayed health effects. At the same time, poverty is relative—too many regions in the world are considered poor compared to Western lifestyles and resources—and historically specific. Even more important, material considerations and lack of resources are not necessarily antithetical to public discussion of risks. As mentioned above, Chernobyl was at its most visible in the very last years of the Soviet Union, which were economically dire times. Indeed, Adriana Petryna has demonstrated that, in Ukraine, Chernobyl-related injuries were not ignored even as the state became poorer with independence and the transition to a market economy. The affected populations received monetary compensation for their Chernobyl-related injuries. Chernobyl thus came to serve as an economic resource for the affected populations in Ukraine, helping them survive (Petryna uses the term "biological citizenship" for the process in which the affected individuals claim protection and compensation from the state based on their injured health). But she also notes that things started changing toward the end of her research in Ukraine.[11]

The period of greater visibility for Chernobyl's effects was followed by the gradual production of invisibility in both Ukraine and Belarus, where the national intelligentsia and local scientists had aspired to document and demonstrate the scope of the fallout and its consequences. The circumstances in the two countries were different, and the trajectories of changes were not exactly parallel, even though the period of visibility in both cases stemmed from a reaction to the same past—the Soviet secrecy around the accident, along with the downplaying of the accident's scope and the nearly criminal negligence of protection for the affected people. Reactions in both the Ukrainian and Belarusian republics were fueled by political transformations in the last years of the Soviet Union.

Yet there were significant differences, including in what Chernobyl meant for the two countries after the collapse of the Soviet Union. Belarus, a smaller country with no significant history of radiological research, faced a greater scope of contamination, but without the physical presence of the Chernobyl power plant that would act as a source of visibilization of the accident even as it posed a whole set of other significant challenges.

Chernobyl is not the only case in which the consequences of a catastrophe have been rendered invisible. Petryna has commented on the difference between her fieldwork, conducted during a period of relative visibility and openness in Ukraine, and "the closing down of fieldwork frontiers" faced by anthropologists studying risks-related justice in more recent contexts.[12] Nor are Western democracies immune to the "disappearance" of hazards. Recent studies describe the lack of public and official recognition, or the delayed recognition, of such risks as tobacco smoke, global warming, air pollution, and hormone disruptors in the food supply. In a particularly relevant study, Michelle Murphy has described the production of imperceptibility—the inability to detect chemical hazards—using the example of sick building syndrome.[13] Science, which we trust to provide us with relevant knowledge, is increasingly an object of outside political pressures, especially from interested industries creating areas of uncertainty and ignorance.[14] Indeed, it is not the disappearance of Chernobyl, but the surge of its visibility in the last years of the Soviet Union that might be atypical.

The Production of Invisibility: Articulation and Infrastructural Conditions

I will refer to the process of defining the scope and character of radiation danger and its actual effects, along with how to make them observable, as *articulation*. I do not isolate scientific research or representations of post-Chernobyl radiation; instead, I consider them part of an overall picture of the layers of practices and processes that affect public visibility of the post-Chernobyl contamination and its health effects.

In practice, several dimensions of representing radiation and its effects are particularly important to their visibility. The first dimension is the temporal and spatial scope of the consequences: Where are the consequences of Chernobyl, when do they appear, and how long are they going to last? As noted above, the consequences of Chernobyl in Belarus (defined by, among other things, formal criteria of what is considered dangerous) came in waves: the scope of recognized consequences fluctuated from very limited during the first several years after the accident, to vast and long-lasting in the two years before the collapse of the Soviet Union, to gradually shrinking in the mid-1990s. The areas recognized as contaminated were later scaled down to, as one of my interviewees put it, "about the size of an airport" (see chapters 3 and 4).

Second, the visibility of Chernobyl depends on how the problem is identified and framed: What kinds of consequences did Chernobyl have? For example, starting from a particular historical point, Chernobyl in Belarus has become almost exclusively an economic problem.

Third, the visibility of Chernobyl depends on what expert and lay practices are used to make it observable: How does one observe Chernobyl's consequences? Scientific research methods that are less sensitive to the complexity of the situation on the ground might allow for ignoring all or most of the "incongruous" data as anecdotal or unverifiable, thus ignoring what could be Chernobyl's health effects (see chapters 5 and 6).

The answers to these questions—where, when, and what are the consequences, and how to make them observable—often involve complex social negotiations, power struggles, and technoscientific work.

Articulation is both a discursive and a material process.[15] Representations of Chernobyl-associated radiation and its effects are shaped in the course of interactions between different social perspectives. At the same time, articulation requires appropriate tools and material conditions organized in a particular way—it requires infrastructural conditions. Addressing these as two separate aspects of the same process allows for some important distinctions. Articulations as discursive definitions of danger are not absolute (accurate or not accurate) but relative and dialogical: they make radiation *more* or *less* visible. Infrastructural conditions define the limitations of this process, and the production of invisibility might become irreversible if the infrastructural conditions required for articulating the presence of radiation and its connection to health effects are disrupted.

Thus, in some respects, articulation—and consequently, the production of (in)visibility—is dialogical and relative, specific to particular dialogues and contexts.[16] The same proposed thresholds, for example, might expand the visibility of a hazard in one context but limit it in different circumstances. Dialogues are always local—with unique contexts, ranges of topics, and perspectives.[17] Perspectives on particular issues are always part of a dialogue: they are grounded in certain social positions (and can thus be described as situated viewpoints), but they are also developed in the course of interaction with other perspectives and reflect the history of this interaction.[18] The theoretical significance of the dialogical approach is precisely in this acknowledgment of the coshaping of different perspectives, as well as the situated and embodied character of interpretations.

This approach also reminds us that dialogues are constantly evolving. New perspectives on Chernobyl continue to appear even two decades after the accident. Some older themes can be overshadowed or (temporarily) disappear, while other perspectives can reemerge. Yet new interpretations do not fully replace the existing ones; they supplement them. (Even with the most radical historical transformations, the whole discourse cannot be assumed to change and be transformed into a new discourse.)

There are two ways of interpreting what is being articulated, and I will use *articulation* in its more inclusive sense. On the one hand, articulation refers to defining the hazard—that is, learning the difference between the presence and the absence of effects (which requires appropriate tools and conditions). On the other hand, articulation extends beyond defining the hazard and includes explicating the work that has to be done to mitigate it, along with the conditions and resources available for this work. In this sense, articulations not only define radiation risks and effects but also account for existing infrastructural resources and shape future ones.

At the same time the very possibility of articulation often depends on the existence of adequate infrastructural resources (e.g., meters for internal and external radiation, stable databases, or even just public spaces for articulation). By infrastructures I specifically mean the information systems and equipment that support the practices of articulating Chernobyl-related effects by expert and nonexpert communities. These infrastructures are embedded in existing institutional arrangements, and they invisibly support research tasks and the accumulation of data.[19] Lack of adequate infrastructural conditions might disrupt data collection, preclude areas of analysis, and foreclose opportunities for articulation. Indeed, radiation and its health effects become strictly unknowable precisely at this level of disruption of material conditions.[20] With the disruption of the material conditions for research and accumulation of data, the invisibility of radiation health effects is no longer relative, and the construction of ignorance comes to be irreversible.

Perhaps we do not always need much public visibility of a hazard, in the sense of rampant public discussion about it, as long as its visibility, the recognition of the hazard, is built into infrastructures of radiation protection and there are adequate mechanisms for research and decision making. This would not address all questions of social memory about the scope of the accident, but it invites us to consider what protection mechanisms there are and what view of radiation risks these mechanisms enact. From this perspective, the ultimate problem is not that Chernobyl has disappeared from public view but whether and to what extent it has worked itself into various kinds of infrastructures.

In the post-Chernobyl context, the systematic disruption of the infrastructural conditions necessary for knowledge production and the limited articulation opportunities for groups that could potentially resist the dominant articulations (e.g., laypeople affected by the Chernobyl fallout, civic movements, political opposition groups, or local scientists) are the two main obstacles to making radiation health effects publicly visible. Among the most troubling signs of limited articulation opportunities is the growing lack of Chernobyl experts in Belarus—that is, scientists who would publicly claim expertise in Chernobyl-related research (see chapter 6).

Our interpretation of invisible hazards is also likely to build on past discourses. What was deeply lacking in Belarus as part of the Soviet Union was any history of public discussion about invisible hazards, along the lines of, for example, Rachel Carson's *Silent Spring*.

Invisibility and Power

So far this introduction has focused on defining the production of invisibility. The rest of the book describes actual practices and conditions that can enhance the public visibility of hazards or, on the contrary, make them less visible. This section discusses the scale of the production of invisibility: To what extent can imperceptible hazards be made publicly invisible and unobservable? To what extent does the production of invisibility affect what we ultimately know? Would not a massive wave of health effects still somehow reveal itself? When asked this last question, a Belarusian physician who used to work with the affected populations (until political changes in the Chernobyl health-care administration system described in chapter 6) replied, "It depends on who's in charge, doesn't it?" The production of invisibility is a function of power relations. At the same time, there is a certain "resistance from reality" that makes this process more or less constrained and effortful.

Relations of power and the interests of those in power matter; even easily diagnosed and observable health effects can be made publicly invisible under conditions of institutional or political secrecy. Furthermore, what is "easily observable" reflects certain properties of the hazard and its health effects—but always in relation to observational practices and established knowledge (which also represents the history of articulations, in this case). Consider, for example, what would happen if the international nuclear industry, local governments of the affected countries, and various groups of experts were as interested in identifying and keeping track of Chernobyl effects as, for example, the U.S. banking industry is in identifying and keeping track of all credit-related activities. If Chernobyl-related data collection infrastructures were as extensive, and the underlying categories as consistent, then one might expect data losses to be minimal as well, thus ensuring a greater scope of registered effects.

If we define the reality of Chernobyl's radiation effects along the lines of Latour's definition of reality as that which resists—in the sense of resisting arbitrary statements and productions—the reality of the accident's consequences is in a kind of interaction with infrastructural conditions and history and opportunities for articulation.[21] One example is thyroid cancer in children, the only Chernobyl-related radiation health consequence acknowledged by the international nuclear experts—who recognized it only after several years of discrediting the data of local scientists and their expertise. For the local scientists, it simply "could no longer be denied": the rate of this notably rare condition was demonstrated to have greatly increased (from an average of one case per year in Belarus).[22] The disease was occurring in a population not usually affected by it—children—and the rise in incidences was concentrated in the most contaminated areas.[23] There were also an established register, a dedicated and experienced researcher and his team (politically savvy enough to establish good protocols and promote their data), and enough of a preexisting scientific consensus about thyroid sensitivity to radiation exposure.

Such a good match between infrastructural conditions, the history of articulation, and the properties of health effects does not happen for most other health problems. Given the great effort that international nuclear organizations put into disregarding the views that would expand our conception of radiation health effects (and how this reflects on the infrastructural conditions and possibilities for knowledge production in Belarus), other health conditions appear not as pronounced. Potentially radiation-induced conditions are often not specific to radiation exposure; they may also appear after a significant time delay, which makes it easier to question their relationship to radiation exposure and attribute them to other factors.[24]

The discussion above suggests that certain properties of the phenomenon, along with infrastructural and other factors constraining articulation, prevent radiation effects from becoming more publicly visible, scientifically observable, and knowable. Perhaps the most important property of radiation in terms of this discussion is its imperceptibility with the unaided senses. A phenomenon does not have to be imperceptible to be made publicly invisible, yet imperceptibility certainly makes it easier to erase public awareness of that phenomenon. The question of perceptibility is thus again a question of power. In some contexts, even obtrusive hazards might be ignored. One might recall hazards that plague socially disadvantaged groups and require large-scale infrastructural solutions. The aftermath of Hurricane Katrina, for example, prompted Susan Leigh Star to ask, "What does it mean for something to be in plain sight and also invisible?"[25] Similarly, the production of the invisibility of Chernobyl's consequences has much to do with the marginality of the affected populations, who are mostly poor, rural residents. International "invisibility" of the country, Belarus, (a rather small, Second-going-on-Third World state) might be an additional factor. Chernobyl-related problems in Belarus are then doubly invisible; from the international perspective, Chernobyl is typically associated with Ukraine, where the Chernobyl nuclear power plant is located.

The emphasis on the production of invisibility as a function of power relations should not be interpreted to mean that the production of invisibility is always an unjust process with socially unacceptable outcomes.[26] The production of invisibility is inevitable to the extent that paying attention to something also always implies not paying attention to something else.[27] Furthermore, Robert Proctor reminds us in his analysis of the social construction of ignorance that, "not all things are worth knowing at all costs."[28] In some cases, what is thought of as dangerous might pose little actual threat, and the production of invisibility might then be a process of learning and a matter of justice. But different hazards do not have equal chances of being made more visible, and precisely that should be the matter of public discussion. How we know what we know, what social mechanisms guarantee that attention is paid in some critical instances, and what ensures that adequate knowledge is produced are key public questions precisely because the production of invisibility is a function of power relations. The particular concerns here include the presence and maintenance of adequate infrastructural conditions, the democratic organization of expertise, and framing of research questions.

Maps: Methodological Considerations and Overview of the Book

I originally set out to study different interpretations of Chernobyl's consequences, but my initial fieldwork was complicated by the fact that I remained uncertain about the actual extent of these consequences. Positioned far from the affected territories and faced with conflicting reports, I was not sure whether there was a real problem behind the representations I set out to study. If radiation from Chernobyl had no significant consequences, then what was the purpose of talking to laypeople about it? Or if the radiation had decreased dramatically since the period immediately after the accident, why should people living in the contaminated territories still care about radiation risks, and why would there be any Chernobyl reports in mass media?

There was a large nuclear accident, but the accident was a couple of decades ago. There could have been thousands dead and millions sick, or it could have been more or less a "myth." Consequently, I chose to rely on what minimal scientific-administrative consensus there was: I bought a map of the current scope of the country's contamination (based on contamination with Cesium-137). The map had enough colored area indicating contamination—about 21 percent of the whole territory, according to the 2001 map—to suggest that the problem was still there.[29] It is worth noting that *certainty* here is again related to *visibility*, or at least knowing what to look at and point at.

The shifting contours of Chernobyl had other implications for conducting research: the kinds of people I would want to talk to depended on how one defined the problem. The categories of Chernobyl-related groups are themselves shaped by the same processes and social interactions that have shaped public understanding of the accident and its consequences. In order to learn the range of perspectives on the consequences of Chernobyl, I began by interviewing laypeople living in the areas officially defined as contaminated. I also interviewed Chernobyl experts—scientists, physicians, government administrators, members of international projects, and members of nongovernmental organizations—whose professional activities are related to Chernobyl knowledge production practices. My selection of "expert" interviewees, collection of document sources, and data analysis were guided by grounded theory methodology; I was seeking perspectives that either further explained or, even more important, contradicted theoretical concepts that started emerging early in this research. Data collection for this project began in 2003, with the main ethnographic part of the research conducted in 2005, when I interviewed local experts and residents of the contaminated areas (see the appendix for further explanation). I continued collecting data and returning to Belarus in the subsequent years. A series of follow-up interviews with local experts was conducted in 2010–2012.

As I was conducting interviews and collecting documents, it became apparent that to interpret the current state of knowledge production practices, it was necessary to understand the history of transformations of public discourses on the topic, especially the official discourse. Systematic analysis of 20 years of media coverage (including government and oppositional publications) helped reconstruct these transformations and provide a more comprehensive perspective. Documenting changes in the official discourse, in particular, created a kind of backbone for my analysis; it helped reconstruct the historical waves of Chernobyl's invisibility. This, in turn, helped construct the interpretative framework for other types of data, including data from the interviews and observations.

The production of invisibility of Chernobyl's consequences in Belarus has been a cumulative, layered process in which the layers correspond to relationships among various interest groups and often reflect power disbalances among them. The key set of relationships is at the top, which includes the Belarusian government, local scientists, and international organizations and their experts. But in describing the production of invisibility, I begin by explaining why the affected public is silently complacent with these processes (chapters 1 and 2).

I then outline the historical trajectory of the transformation of Chernobyl's consequences—the waves of their (in)visibility—as illustrated by the discourse in the official and oppositional media (chapter 3).

The second half of the book (chapters 4–6) explains how the waves of invisibility of Chernobyl were shaped by Chernobyl-related interests of international organizations, state management of research infrastructures, the resistance of some local scientists, and political battles over the formal principles of radiation protection.

Throughout the book, articulation and its infrastructural conditions are emphasized as the main aspects in the production of (in)visibility of Chernobyl radiation and its health effects. Some chapters (2, 4, and 6) underscore the problems, paradoxes, and double binds of infrastructural solutions. Other chapters (1, 3, and 5) are more concerned with the articulation processes, the kinds of articulations put forward, and the opportunities for lay and expert articulations.

Chapter 1 thus argues that even for people living with radiation, experiencing it and developing knowledge about its effects depends on how the danger is articulated, which in turn is shaped by opportunities for articulation. In the Belarusian post-Chernobyl context, much-needed dialogical opportunities for articulation are strikingly limited. As a result, the affected populations often rely on readily available administrative discourse (rather than discourses based on science or laypeople's own collective experiences) to define the scope of radiation danger and its health effects. Based on this analysis, I argue that the affected populations cannot be assumed to be the most risk-conscious.

Chapter 2 discusses the multiplicity of lay perspectives on risks, partly focusing on those affected individuals, families, and communities who have accumulated the highest doses yet are the most resistant to making radiation risks more observable. The residents of the contaminated areas show different levels of concern about radiation and, correspondingly, there is a range of accumulated internal doses, which people acquire by consuming radioactive materials with food. Years after the accident, individual doses are a matter of individuals' own choices and behaviors; for example, people can avoid significant exposure by not consuming wild berries or mushrooms.

But even though people "make their own doses," they do so not in circumstances of their choosing. The physical properties of the distribution of radionuclides in the environment come to be aligned with the system of socioeconomic privilege, so that the least socially advantaged groups are exposed to greater radiation doses. Furthermore, some local populations resist articulation processes that could make contamination more observable simply because mitigating this contamination would then require nearly constant work that far exceeds their individual or family resources. Finally, interpretations of risk are also affected by individuals' own trajectories with respect to the hazard—that is, the extent to which radiation exposure remains a current problem or has happened in the past.

Chapter 3 describes the waves of (in)visibility of Chernobyl—the different framings and degrees of recognition of the temporal and spatial scope of Chernobyl's consequences—as they appear in the official discourse in mass media during the first 20 years after the accident: secrecy and silence in 1986–1989, an explosion of media attention to Chernobyl in the last years of the Soviet Union and the first half of the 1990s, and a gradual transformation and narrowing down of the scope of Chernobyl problems thereafter. This chapter provides a detailed historical context for these transformations, and it also considers the double-edged discursive phenomenon of *hypervisibility*: exaggerated, dramatic, and stereotypical portrayals of radiation effects (such as images of bald children with cancer) that were employed by the political opposition in the late 1990s to draw attention to the disappearance of Chernobyl's consequences in the official discourse and counteract the government strategies of rehabilitating the affected territories.

The second half of the book describes the waves of (in)visibility of Chernobyl as reflected in the formal representations of Chernobyl's consequences, the organization of Chernobyl-related scientific research, and the international constraints affecting knowledge production practices in Belarus. The local government has been confronted with resistance from the international experts who claim minimal Chernobyl radiological effects. It has also been faced with the scope of the consequences that—if defined to the maximum—far exceeds the capacities of the state, especially given the lack of international assistance. Local scientists, in turn, have been dealing with pressure from international nuclear experts and an ideologically over-bearing government. This set of relationships has been unfolding differently in different periods, leading to fluctuations in Chernobyl's visibility and, ultimately, to the disappearance of almost any socially and bureaucratically recognized Chernobyl contamination and health effects.

Chapter 4 examines the succession of approaches that defined Belarusian radiation protection efforts. The scope of radiation danger recognized by the government and the official view on required protection measures depend on the adopted radiation protection concept. It is developed by scientists, and like most formal representations (e.g., thresholds or standards), it appears neutral and objective. In Belarus, this radiation protection concept was redefined several times, and each time the scope of the recognized radiological contamination and its risks shrank or expanded radically.

This chapter explains the politics of formal representations—that is, the ways in which they make radiation danger more visible or less visible. Specifically, I argue that the production of invisibility depends on how formal representations are aligned or misaligned with the empirical complexity of radiological contamination and what could be measured in practice. The work of alignment is fundamental to making radiation and its effects visible. Often it can be done only by experts, from their particular bureaucratic and technoscientific positions, which highlights the importance of accountability, the democratic organization of expertise, and the value of oppositional experts.

Chapter 5 considers the history of the Chernobyl-related research of the United Nations. Experts from the International Atomic Energy Agency and the World Health Organization vehemently supported the Soviet scientists in their denial of any significant consequences of the accident. This position remained unchanged for two and a half decades. The chapter describes the research and rhetorical strategies used by the UN nuclear experts to minimize public visibility of Chernobyl health effects.

This chapter also considers how and why the Belarusian government gradually came to concur with the United Nations on the issues of Chernobyl in the early years of this century, after a period of disagreement and limited UN assistance. The new strategy supported by both sides reframed Chernobyl as an economic problem and a problem of sustainable development. This framing not only provided the grounds for cooperation but also implicitly reasserted the same minimizing view of Chernobyl health effects and allowed for new approaches to making radiation danger less publicly visible, both in Belarus and internationally.

Chapter 6 describes national research efforts, which have been affected by the peripheral position of Belarusian science (first within the Soviet system of Moscow-centered science and then in relationship to international nuclear expertise) as well as by the economic and political interests of the Belarusian government. These conditions led to the production of invisibility in two direct and immediate ways. First, knowledge production is subverted through the reframing of research so that the radiation factor disappears as an object of inquiry, along with local experts who would claim expertise in Chernobyl radiation health effects. Second, politically induced infrastructural disruptions to data collection and analysis create the conditions for research relying on theoretically, rather than empirically, driven approaches, and this bias supports minimizing the scope of Chernobyl-related health effects.

Making radiation risks and health effects visible is work. The production of visibility particularly depends on the work of articulating and of creating and maintaining adequate infrastructural conditions. Furthermore, the result of the work of articulation is more work—necessary to manage the chronic problem of radiological contamination (which again most likely requires large infrastructural solutions). This book considers what work could not be done in Belarus to make the radiation risks and effects visible—and what work was done to make them invisible.

1 Articulating the Signs of Danger

A local member of an international Chernobyl project, a young man who grew up in one of the more contaminated areas, asked me why foreigners were interested in solving the problems of Chernobyl. "It is mostly foreigners who are passionate about Chernobyl problems, and not the local people," he observed. The attitudes of visitors might change, however, as they witness the actual circumstances in the contaminated areas. Another local resident and member of the same Chernobyl project argued that foreigners "come here and see that everything is normal. Radiation is scary only the first time you go to the [Chernobyl] regions and see abandoned houses there." The first few times foreigners visit, according to a member of a local Chernobyl-related nongovernmental organization (NGO), they show more concern and take more precautions. Some international humanitarian teams bring their own food, water supply, and even a chef; they might also eat in restaurants, since the food there has to pass state radiation inspection. This local resident commented that foreigners who come to the contaminated areas "take many measures, and those who live there permanently ... " He waved his hand, indicating nothing, no precautions.

Those who should worry most, or at least more, often appear to be the least concerned; the experience of living with increased radiation danger does not necessarily bring out more anxiety. In at least some cases, it has the opposite effect.[1] Svetlana Alexievich, a Belarusian journalist who has described the aftermath of Chernobyl in Belarus, observes that Chernobyl has not been experienced in the same way everywhere and that what is remarkable about the residents of the most affected areas is the indifference with which they talk about it.[2] Although care should be taken not to suggest that all residents of the contaminated areas are indifferent to the danger, this chapter focuses on how imperceptible hazards are experienced—and why this experience might result in less, rather than more, vocal concern about the hazard.

How are the experiences of a hazard shaped if the hazard itself is completely imperceptible with the unaided senses and most of its health effects are delayed in time? How we interpret the nature of this experience affects theoretical perspectives on what expertise laypeople are assumed to have, how we position their expertise in relation to that of experts, and when laypeople's experiences are assumed to contribute something valuable to expert assessments. As discussed in the introduction, Ulrich Beck warned us that laypeople lose the autonomy of their judgment when traditional cultural mechanisms cannot identify and make sense of invisible hazards: to recognize the risks of radiation, people must not trust their senses, which register nothing. Laypeople become dependent on scientific and administrative practices, and on the media, for identifying risks. We need the "sensory organs of science" to make hazards visible.[3] At the same time, the affected lay populations clearly do have extensive experience of the political and socioeconomic aspects of living with increased levels of radiation. People are acutely aware of social context and the power dynamic inherent in the advice given by scientists and administrators. Laypeople doubt or even ignore such advice when its politics threatens their identities and seems instead to promote the interests of expert institutions.[4] My question, however, is specifically about lay interpretations of radiation danger and the nature of various experiences (not limited to interactions with scientists) that lead affected groups to arrive at these interpretations.

The imperceptibility of radiation means not only that the contaminated environment and food look exactly like uncontaminated ones but also that there might be no readily available categories to help the affected communities observe and make sense of the situation. In other words, there might be no spontaneous, commonsense interpretations for what the hazard is, what its effects might be, or when and how they might be observed. The availability of naturalized categories that give meaning to perceived phenomena is fundamental to experiencing them.[5] In the case of imperceptible hazards, both foundations of experiencing—sensory perception and naturalized, readily available explanatory categories—might be missing, at least at first. Aleksievich observes that many Belarusian residents she interviewed in the early 1990s mentioned not being able to find words to describe what they saw and felt after Chernobyl. She comments, "Something occurred for which we do not yet have a conceptualization, or analogies, or experience, something to which our vision and hearing, even our vocabulary, is not adapted. Our entire inner instrument is tuned to see, hear or touch. But none of that is possible."[6]

Making sense of imperceptible environmental contamination differs from assessing "conventional sudden-impact events (either natural or manmade)," in which commonsense interpretations emerge spontaneously and there is little doubt about the relationship between cause and effects, even if individual reactions to these hazards may vary greatly. Martha Fowlkes and Patricia Miller describe the reactions of Love Canal residents to the chemical contamination of their community: "Each family found itself in an unusual and difficult position of having to evolve its own definition of the significance of the chemicals. Facing either the possibility or desirability of relocation, families were *required to articulate coherent perspectives* about the actual or potential implications of the chemicals on their wellbeing" (emphasis added).[7]

The imperceptibility of radiation with the unaided senses thus means that radiation danger and its possible health effects might not be spontaneously obvious to those who experience them. The signs of radiation danger and its connection to actual or potential health effects have to be articulated: identified, explicated, and established as such. Radiation is not visible or observable to laypeople living in the contaminated areas without this work of articulation.

These articulations certainly might become commonsense over the course of time; we need not assume that laypeople remain culturally blind to imperceptible risks forever and under all circumstances. New hazards, under new historical circumstances, might also require rearticulation. At the same time, the post-Chernobyl contaminated communities might have been particularly unprepared for this task of articulation, not least because of the lack of previous exposure to organized environmental activism— which would have drawn public attention to other imperceptible hazards such as chemical toxins—and because of the general suppression of civic society under the Soviet regime. My question then is how people develop ways to identify, interpret, and imagine these imperceptible hazards—and what conditions are necessary for these articulations to take place.

This chapter's approach to what is articulated and how builds on Bruno Latour's discussion of articulation as the process by which one's body is "learning to be affected," or learning to distinguish when being affected in different ways. Latour uses an example of training to become a "nose" for the perfume industry: one's ability to differentiate smells is developed in the course of a training session, with a kit of contrasting odors and a teacher. At the beginning, the student is inarticulate and reacts in the same way to different sample odors. In the course of the session, however, the body learns to be affected and to tell the difference between odors. Essential to this process is mediation by the kit of contrasting odors and access to the teacher who has benefited from the accumulated body of knowledge in chemistry. Although Latour does not emphasize it, equally significant is the role of various categories that define odors and help the trainee make sense of her sensory experience. This approach to learning to be affected as a mediated, interactive process applies to laypeople as much as experts; experiences of one's "lived-in body" similarly depend on mediation by material artifacts.[8] The emphasis on the mediated, interactive nature of the process allows us to analyze historically and contextually specific practices that shape lay experiences of imperceptible hazards.[9]

We should, however, keep in mind several differences between the context of imperceptible hazards and Latour's example of training a "nose," a case in which articulation is essentially the education of the senses. Unlike perfume, radiation is imperceptible, which makes lay articulations of radiation danger even more dependent on material artifacts: radiation has to be made observable with radiation counters and other tools and representations. Although bodily experience is not absent with radiation exposure (some people get sick), it is delayed, which makes it more difficult to articulate the connection between radiation exposure and the state of one's health.

Lay articulations of radiation danger also happen in contexts far broader and more diverse than the narrow laboratory setting of nose training. In Latour's example, learning to be affected takes place in a tutorial with a goal-oriented setup, a teacher with a clearly defined area of expertise, and through the use of established tools, categories, and representations. Lay articulations of radiation danger are not limited to such established contexts, tools, and representations. We therefore must ask (and this will be the central issue for the rest of this chapter): What sorts of mediation setups do laypeople use to note increased levels of radiation and their effects? What tools and expertise do they use (i.e., what are the equivalents of the odor kit and the teacher's expertise in the perfume training situation)? What categories do the affected communities use to describe the situation, and how do they arrive at these categories? What kinds of interactions—what other perspectives and contexts—shape the process of articulation?

The rest of this chapter considers contexts in which people's activities are related to Chernobyl and in which, through interactions associated with these activities, laypeople can articulate the scope and significance of local contamination and its radiation effects, as well as what can be done about them. I consider what interpretative frameworks, or lenses, these contexts allow for. Explicitly science-based articulations might be the most effective in making radiation observable and publicly visible, but it is by no means the only or even the dominant way of representing radiation danger and its effects in Chernobyl-affected areas. Consequently, the examples below also illustrate how articulation takes place in administrative contexts. The availability of spaces for articulation, the nature of interaction and mediation there affect how lay communities articulate radiation danger they might be facing: what connections (environmental, health, socioeconomic, historical, etc.) are made visible and what connections are never articulated and thus become de facto nonexistent. In short, I consider what is made visible and what is excluded and ignored.

The discussion below is based on observations made and interviews conducted during trips into some of the affected districts—trips that I made with radiologists from the Institute of Radiation Safety "Belrad" (the only independent Belarusian NGO working to identify the scope of radiation danger and provide comprehensive radiological protection for the population), and then with a team from the international CORE program ("Cooperation for Rehabilitation of the living conditions in the Chernobyl-affected areas in Belarus").[10] The Belrad team focused on measuring the levels of internal contamination (i.e., radionuclides consumed with food) in children in a local school. The trips with the CORE team were to other, more significantly contaminated districts. The team encouraged local initiatives for its socioeconomic rehabilitation projects and held introductory meetings in three villages (see the appendix).

Most of the people I talked to had been living with increased radiation risks for close to 20 years, ever since these areas became contaminated. They remembered earlier government policies on Chernobyl and were well aware that the government policy since the mid-1990s sought to "rehabilitate" the affected territories and "normalize" people's lives there, especially since the local administrative actions echoed the official discourse in the media (see chapter 3). I quickly came to doubt my initial assumption that the affected communities would necessarily want to learn more about the risks in their daily lives. As will become clear, at least some groups within these communities might be motivated to ignore particular dangers.[11] Chapter 2 will examine the diversity of actual behaviors in the affected communities faced with the problem of chronic, pervasive contamination, the mitigation of which requires continuous and demanding efforts.

The examples in the next section are based on my trip with members of Belrad's Whole-Body Counter Laboratory. I will call the village we visited Selo. A smaller nearby village, which was resettled in the early 1990s because of the greater levels of contamination there, will be referred to as Otseleno. Chernobyl fallout did not cover the area evenly, and both villages are inside spots of contamination surrounded by a larger, relatively uncontaminated area.

Interpreting Radiation Danger in the Context of Radiological Testing

On this trip, a radiologist and a driver from Belrad brought a whole-body counter (WBC) to measure internal radiation in children from a small local school in Selo, a community of about 900 residents. The school stood at the center of the community, next to a church. The building next door housed a *detski sad*, the equivalent of a preschool. The school, including the preschool, had about 180 children at the time, some of whom came from adjacent villages. A spacious yard surrounded the two buildings. The interior of the school appeared modest but also remarkably neat and well cared for, with green plants hanging in front of the windows along the halls and in the classrooms. Most of the children—born many years after the accident, in this area of relatively low contamination—had not been tested before.

The WBC, which measures gamma radiation from elements within the human body and looks like a chair connected to a computer (figure 1.1), was set in one of the classrooms. The head of the school, a young man in his late 30s, and the teachers coordinated the children while Nikita, the radiologist, calibrated the counter. Nikita then invited the children, who were visibly excited about the procedure, to sit in the chair, one at a time. The primary goal of the trip was to test schoolchildren, but anyone was welcome to be tested, and quite a few teachers and staff members were. Adults and children old enough to read were handed a leaflet with the results of their testing noted next to the green-yellow-red scale meant to put these results in perspective (figure 1.1). Inside, the leaflet listed instructions for how to lower internal radiation exposure by treating meat, milk, mushrooms, and other foods.

Figure 1.1   ▶
Cover of the WBC testing leaflet, Belrad's WBC Laboratory.

The picture shows a boy sitting in the chair of the whole-body counter (WBC) and a radiologist looking at the testing results displayed on the computer. The green-yellow-red scale next to the picture provides a comparison scale for the individual number recorded next to the scale (the scale is in becquerels per kilogram).* The numbers "00" handwritten on the cover show that in this case the subject tested did not have internal radiation above the sensitivity level of the meter.

* The scale used by Belrad is much lower than the acceptable limits set by the Belarusian Ministry of Health. Belrad refers to its scale as control thresholds. Their estimation of danger is based on the work of Yuri Bandazhevsky (see chapter 6).

Институт радиационной безопасности «Белрад»

Лаборатория спектрометрии излучения человека

Удельная активность радионуклидов цезия-137 в организме Вашего ребёнка:

70

20

00

0

(Бк/кг)

Фамилия: Кучинская

Имя: Ольга

Отчество: Геннадьевна

Год рождения:

Дата: 24.11.04

The WBC is a particularly powerful tool for making radiation, especially internal radiation, "visible" to laypeople. During a WBC testing session in an institutional setting (school), internal radiation becomes the focus of everyone's activity. As we shall see, other contexts are subsumed; competing explanations (i.e., what is or is not hazardous) and motivations (including economic concerns) are largely ignored. Such an organized and restricted context is exactly what enables people to "see" radiation danger most explicitly.

In a setting like this, radioactivity in people's bodies is made visible with the radiologist's professional expertise and tools, including the WBC and the scale for interpreting the results. Conversations prompted by the testing attempt to make sense of what has been made visible, negotiating causal explanations for doses marked as high or low. This meaning-making process is organized by numbers produced by the WBC as well as the authority of the radiologist who interprets them. And even though a radiologist familiar with the local way of life is generally aware of people's broader concerns (such as the economic situation), references to other contexts largely remain suppressed. In later sections, we will return to other ways of interpreting radiation danger that reflect the concerns of daily life more inclusively, even though radiation danger might not be as easy to observe in these settings.

"Professional Vision" and Making Visible

During a WBC testing, radiation is made visible not only by the counter itself but also by the radiologist, who actively interprets the numbers and makes use of the green-yellow-red scale to explain the relative severity of doses (figure 1.1). Charles Goodwin refers to the ways in which experts organize reality, highlighting what is important and what should be ignored, as "professional vision."[12] The work of the radiologist in this setting is to bring people's attention to what is radiologically important; the rest is left outside the frame, outside the radiologist's focus.

During the testing session in Selo, a number of activities performed by Nikita helped construct the visibility of radiation in that setting. Nikita provided the schoolchildren and teachers with the results of their testing, interpreted the significance of their internal radiation doses, reframed people's past activities in ways that emphasized the sources of their internal contamination, and gave advice on how to reduce this contamination. For example, Nikita asked a small boy whose number was in the red zone of the scale whether he had been eating mushrooms (gathering forest mushrooms is a popular activity, practically a national sport, and wild mushrooms accumulate particularly high levels of radionuclides). The boy replied that he had. Nikita suggested that the boy drink Vitapekt, an apple-based food supplement that absorbs radionuclides. Highlighting causal connections—what leads to greater internal radiation doses and which levels of doses might lead to health problems—is as much a part of making radiation observable in the context of radiological testing as determining the doses themselves is.

Interpretations, judgments, and explanations made by the radiologist were adopted by the children and adults being tested. One boy told his number, which was high on the scale, to his classmates later, and one of them commented, "What are you, irrational [*nerazymny*]?" Children asked one another, "Show me, what radiation have you got?" The teachers and staff made comments like "I'm going to spy around, [see] who's got the highest radiation" or "It's just left to see who's got the highest doses." This kind of discussion, of who had the highest doses and why, took place throughout the time that Nikita made his measurements, class by class, person by person.

A WBC testing provides an occasion for considering causal explanations and indeed for reimagining the role of radiation in one's life. The discussion in Selo went beyond comparing levels. Children and teachers reinterpreted their shared backgrounds *through the lens of the measurement*. In this context and in the course of these discussions, they articulated their views of the scope and nature of local contamination, and they collected their own popular statistics. Explanations for how people got high levels were advanced: "She's a milk lover," or when a boy's WBC results were much higher than the other children's, a teacher commented to her coworkers that his father was a forester. The school nurse got up from the WBC chair after learning her low number and immediately offered an explanation, "I don't eat meat and I don't eat mushrooms." Older children discussed the measurements enthusiastically:

"I have the most radiation," one child said, "It's because I'm the fattest."
"No, it's because you eat too many mushrooms," another child insisted.
"I don't eat mushrooms," the first child retorted. "I don't like them."

The context of radiological testing provokes recollections and negotiations of previously received advice and information. The setting provides an occasion for individuals to reexamine explanations for internal radiation in the light of their own testing results and in the light of potentially related individual or community experiences; the validity of these explanations can be reinforced by the authority of the radiologist. In this context, the affected individuals are adopting or at least considering the professional vision of the radiologist. For example, as the Belrad team was packing to leave, a female custodian commented to me about the "boy with the high number":

A boy from the second grade has got one hundred and eighty and something. [He is] just in the second grade.... His mother works as a cook in the school. We asked her if that's from mushrooms. She said no, otherwise she'd have it [a higher radiation dose], too.... But where from, if not from mushrooms? They have a load of them, and they are from that area [the particularly contaminated spot around Otseleno].... No, she does not live there now, but her parents used to live there, and she knows the forests there very well, so she has loads of mushrooms every year.

As might be obvious from the quotes above, people in Selo did, over-all, show considerable levels of knowledge about their radiological situation. For example, an older teacher with a particularly low number thanked Nikita for doing the testing and described her practices, all reflecting past radiological advice: she was not eating mushrooms, was not using ashes as fertilizer for her garden plot, and was not burning wood (she was using gas and briquettes instead). But the radiological advice was not accepted uncritically; the same woman called one piece of advice in the leaflet "nonsense": it suggested that boiling meat in salty water reduces the level of radiation in meat. The woman countered, "Salt is also not good for you. I am used to eating without salt."

Radiological Testing and Broader Contexts

The comments made by the teachers and schoolchildren also reflect their uneasy relationship to the restricted context of testing and the narrow scope of radiological explanations, both of which are rather divorced from the concerns of daily living. The conversations illustrate that people did not necessarily pay attention to details that could be considered scientifically important. One such detail was the difference between external radiation exposure and internal absorption of radioactive substances, a classic issue in the public understanding of science.[13] In some cases, the flawed logic was corrected by one's interlocutors.

Teacher 1: I will have high radiation; I live close to the forest.
Teacher 2: It is not going to affect it, if you live by the forest.
Teacher 1: And I drink a lot of milk and I eat meat. It is not going to be good.

But often the comments were simply imprecise, making connections in rather sweeping ways, as in the following comment about a low internal contamination dose: "Well, you are the cleanest because you live in Selo" (rather than in one of the smaller neighboring villages). When another woman was said to have a higher dose, others commented, "She lives on the street with the highest level of radiation." These observers recalled information from radiological testing of the area done years ago, but they also did not differentiate exact causal connections: a reference to living on the most contaminated street might incorrectly refer to external radiation exposure, or it might imply that garden soil on that street was more contaminated (and consuming food grown in contaminated soil can lead to increased internal doses). This typical lack of the differentiation of external and internal exposure was obvious in a comment made by the school librarian while she was waiting for her WBC test results: "Okay, let's see where I have been wandering." Similarly, the head of the high school (*zauch starshih klassov*) was puzzled about the boy with the highest dose: "It's strange; he was born in Grodno region" (which is considered "clean"). This is not necessarily a problem of the assimilation of radiological knowledge: making precise causal distinctions, though important from the radiologist's perspective, might just not matter as much in people's daily lives.

One might observe other simplifications, arguably made for the purposes of translating radiological data into categories of local practice. For example, local areas with relatively higher contamination or foods that are known to accumulate radiation might come to practically *objectify* radiation. Radiation then appears to be completely and almost perfectly contained within these places and objects, and, to quote a woman critical of this practice, radiation is then "all in mushrooms and never in potatoes." Indeed, everybody I talked to in Selo referred to Otseleno as "contaminated" and the area in the opposite direction as "clean." The following description given by a local farmer was typical: "Over there [toward a resettled village], it's really bad.... Everybody knows that the most radiation is there. All their houses were resettled elsewhere. And over there [a different neighboring village], it's clean."[14]

To take radiation-related precautions, then, is to avoid particular places and objects—for instance, not to collect mushrooms or to collect them only in the "clean" areas. The head of the high school commented, "We don't gather them here, we go to the other side—it's clean there." The residents of other villages had a similarly strong idea about where "clean" and "contaminated" spots were around their villages—even though, according to Nikita, "typically, nobody even looks where it's clean land [or] contaminated land."[15]

As noted above, the context of radiological testing allowed for discussions of what foods tended to accumulate radionuclides and what practices were radiologically unsafe. Yet most of the locals' concerns for daily living were excluded and appeared irrelevant in the context of the WBC testing. There was not much explicit discussion of, for example, the relationship between people's socioeconomic circumstances and their doses, even though this relationship was noted by Belrad radiologists outside the context of testing. Nikita's own awareness allowed for some tacit consideration of these broader issues even within the context of the radiological testing. This tacit recognition of all things excluded shaped Nikita's advice to the locals; it was made explicit when Nikita made comments to me, an outsider. For example, Nikita did not advise the locals to avoid eating mushrooms (the most significant source of internal contamination) altogether, and he explained to me that mushrooms are a much more significant portion of the local people's diet compared to the diet of people in the cities.

At the same time, from the perspective of at least some locals, other concerns undermined the relevance of radiological advice (see chapter 2). Not everybody welcomed the practice of testing. Although most of the teachers wanted to get tested along with the students, one kindergarten teacher asked Nikita, "What do these measurements do for you?" She looked rather unimpressed with the explanation and declined an offer to get tested herself: "I don't want to know." According to Nikita, this happened all the time: "People just think that the less you know, the better you sleep." The principal , who overheard the last comment, replied, "Yes, you sleep better, but not for long." The kindergarten teacher did come back later; she had decided to get tested simply because everybody was doing it.

To summarize: The context of the WBC testing (and not just the counter itself) helps laypeople "see" their internal contamination. It might also help to articulate causal relationships among internal doses, one's daily practices, and the contamination of the local areas. The process of articulation depends on the opportunities that the context of testing provides: its mediational setup, people's interactions with the radiologist and his professional vision (i.e., ways of organizing and interpreting the reality of radiological exposure), and, more broadly, the opportunity for some focused communal discussion. In the context of the testing, radiation is made visible as numbers describing internal doses, but there could still be other ways of representing it (such as objectifying radiation as particular foods or areas). There are, however, other contexts and representations that laypeople use to make sense of their radiological circumstances.

Interpreting Radiation Danger in Administrative-Economic Contexts

Scientific discourse and interactions with scientists are not the only, or even the primary, context that has shaped laypeople's perceptions of radiation danger after Chernobyl. Indeed, the kinds of scientific discourse about radiation effects that have reached local residents in Belarus are fragmentary and rather inconsequential for their daily lives. Locals' interpretations have been built on the history of how the problem has been defined: primarily through a series of administrative policies, echoed in the official discourse on Chernobyl in the media. The locals are thus very conscious of the nuances of the official position, and their own interpretations are shaped through their reflection on government policies and official discourse.

The overwhelming, life-shaping power of administrative decisions (outlined in chapter 3) is reflected in the passive verb forms that are common when locals talk about Chernobyl. In these statements, laypeople are the objects of action, not the active agents. Who does the action is not stated directly, but the presence and power of government authorities are such that they do not have to be named, as in the comment "We *got resettled* [from Otseleno]" (*Nas pereselili*).

One consequence of this strong administrative presence is that laypeople define radiation risks based on the administrative policies adopted to mitigate them. According to one senior member of an international Chernobyl project, a man working in the most contaminated areas of Belarus, "People perceive radiation together with socioeconomic questions; you cannot separate one from the other." Nikita, the radiologist, observed that people judge the level of radiation danger based on whether they still receive Chernobyl-related compensations: "contaminated" spots are the areas where people still receive compensation, and the areas where people are not paid anymore are considered "clean." Only from that perspective does the following exchange make sense:

Author: Are people in Selo concerned about radiation at all?
Local resident: We are not paid anything for this radiation. The only thing is free food for children at school.

I repeatedly heard similar statements in Selo—and was puzzled by them. When I told two locals, a farmer and a forester, that I wanted to talk to them about the problem of radiation in Selo, their reply was "What radiation? We are not paid for radiation anymore. And with what we used to be paid, you could not even buy a pack of cigarettes. There is not much radiation here. There is a lot of radiation in Otseleno," a resettled village.

Not just Chernobyl-related compensation (what people used to call *coffin money*), but also other related government efforts are interpreted as signs of the extent of radiological contamination: gasification (creating gas pipe infrastructure to prevent people from burning radioactive wood in their furnaces); free school lunches for children from contaminated areas (so that children are provided with cleaner food products for at least part of the day); and efforts to disseminate information about radiation protection and set up opportunities for testing food. Administrative practices meant to mitigate various effects of radioactive contamination thus come to *constitute the visibility of the problem of radioactive contamination;* they become the signs of the contamination.

This does not necessarily mean that people blindly trust government definitions or the government's ability and willingness to protect them. Rather, this is an indication that radiation has not been articulated enough in other terms. There have been very few adequate opportunities for public discussion and open forums,—that is, spaces where other and more complex articulations (perhaps incorporating radiological assessments) could be developed and adopted by the local residents. The next section will show that even the health problems that come to be associated with radiation effects are shaped through particular administrative policies.

The danger with this administrative constitution of the visibility of radiation is that erasing the administrative signs then erases the problem. When asked about the scope of radiation danger, the residents of Selo often talked about what "used to be" and contrasted it with what was "now." Some replied that "They don't pay us anymore—have not been paying for three years now." The locals mentioned that preserved food used to be brought to the village "at the beginning" (to minimize consumption of locally grown food) and that "they used to be measuring radiation everywhere. There was a map of [local] radioactive contamination." There "used to be" official information, and "it used to be talked about" by various government representatives. The village used to have its own rehabilitation center—funded by United Nations Educational, Scientific, and Cultural Organization (UNESCO)—where people could get their food tested. Radiation protection was explicitly part of the school curriculum. The area saw gasification efforts that were later stopped.

Some commented that the people themselves used to care and worry, but not anymore. They have gotten "used to it" and "don't do anything special for it ... maybe [only] for their children." One schoolteacher brought up what still remains: "And *still* it is considered that there is some radiation here. There are free lunches at school for children [from contaminated areas]." In these comments, signs of radiation are explicitly constructed; they are the result of decisions and actions of the government, which leaves room for doubting government motives for any of the policies. Yet these comments do not uncouple signs of radiological contamination from the government's actions: the stories of what "used to be" place radiation as something in the past and only to some extent in the present.

Compared to other affected areas, Selo has rather mild levels of contamination, but its experience with discontinued administrative programs is typical. Government programs designed to mitigate the effects of radioactive contamination have been decreasing since the mid-1990s. Even when present, programs like monetary compensation, gasification, and free school lunches do not allow for articulating the connections between radiation and health effects. Establishing these connections is hardly possible when the dominant markers of radiation danger are administrative socioeconomic programs.

Socioeconomic programs provide limited contexts for articulating signs of radiation danger, even when they strive to be more comprehensive and are funded by international bodies. For example, CORE maintained a significance presence in some heavily contaminated regions from 2004 to 2008. The program emphasized "sustainable development" of the affected areas through a participatory approach, but it also included a radiological component. The staff of the program talked about establishing a "practical radiological culture," which would include, for example, the use of particular farming techniques to reduce the radiological contamination of produce.[16] As with programs sponsored by the government, the very fact that CORE was a Chernobyl-related project made radiation danger more visible (see chapter 5). As mentioned above, the program focused on economic and other issues of primary importance to the locals. Yet this and other similar sustainable development programs allowed for only very limited discussion of local radiation risks. Indeed, some Belarusian members of the project privately commented on how little difference there was between sustainable development needs in the Chernobyl-affected areas and rural areas anywhere else in Belarus.

I observed only a few references to radiological issues at the meetings that the CORE team held with the local residents in three affected villages, and all the references were made in passing. For example, at one point a group of farmers and two members of the CORE team were discussing various types of produce for sale: "We have checked onions for radiation. Onions had seven to ten becquerels [units of radioactivity] per kilogram. What's the norm? A hundred becquerels per kilogram." The CORE agronomists confirmed that people were interested in reducing radiation levels, but mostly in reference to the produce they were selling. Very few locals were concerned about radiation in the produce they were keeping for their own consumption. Even in the context of food for sale, the meetings presented radiation risks as no more salient than any other ecological issues that might be considered (e.g., nitrates in locally grown produce).

The CORE program's sustainable development approach thus allowed for broader public participation, and the program meetings provided space for discussion. Yet the socioeconomic emphasis of the program did not allow for nuanced articulation of radiation risks and effects, including any meaningful discussion of radiation-related health problems.

Unarticulated Radiation Health Effects

Like radiation itself, radiation health effects have to be articulated and made visible. Most radiation-induced health problems resulting from chronic and so-called low-dose exposure are delayed in time, and their relationship to one's past radiation exposure is not immediately obvious. Articulating radiation health effects might begin with collecting anecdotal data and observing the prevalence and character of particular health problems among local population groups. Phil Brown describes "popular epidemiology" as "the process by which laypersons gather scientific data and other information, and also direct and marshal the knowledge and resources of experts in order to understand the epidemiology of disease."[17] These articulation practices are not independent of scientific expertise or administrative power. What is diagnosed or recognized as a health problem by laypeople, and how, depends on the existing official diagnostic categories, available health infrastructures, and administrative policies and health programs. Just as various lay definitions of the scope of radiation danger reflect their administrative shaping, so do lay articulations of radiation health effects.

What has been officially acknowledged in post-Chernobyl contexts is the link between radiation and "thyroid" in children ("thyroid" has become common shorthand for a number of thyroid problems, including cancer). Authoritative accounts of the links between thyroid conditions and radiation exposure have appeared in the media (see chapter 3). Many children from the contaminated areas have been tested for thyroid pathologies; some have been diagnosed and treated. Local residents could send their children to state and international health recuperation programs for children with thyroid dysfunctions. Consequently, "thyroid" came to be commonly seen as related to radiation. In conversations with the locals, thyroid problems are likely to come up almost instantly. In Selo, three cases of thyroid pathologies were mentioned to me almost immediately, in all three instances by mothers of sick children. The children had been to Italy for "health recuperation"; they could also undergo medical observation in a rehabilitation center in Minsk if they wanted to. Yet the nuances of the relationship between various thyroid dysfunctions and radiation exposure are typically not discussed. For example, thyroid problems are typically interpreted as the result of radioactive iodine accumulating in the gland, but the children in Selo have not been exposed to radioactive iodine from Chernobyl, since it decomposed shortly after the accident itself; these children were born years after that. To some extent, thyroid and its diseases come to objectify radiation health effects in ways similar to how wild mushrooms and berries come to objectify radiological contamination (see also chapter 2).

Articulations of other potentially radiation-related health problems are much less coherent, and their relationship to past radiation exposure is not articulated with certainty. The lack of references to scientific expertise limits the credibility of some of these accounts, such as this August 24, 1990 description in the newspaper *Sovetskaya Byelorussiya* of schoolchildren from heavily contaminated Vetka:

In the first year [after the accident], the throat reacted—everybody got swollen tonsils, angina [sore throat] tormented. In the second [year]—pneumonia[;]there were practically no children who haven't had it. Thyroids got enlarged. In the third [year]—fainting started, nosebleeds, headaches. There are kids who that year had hepatitis three times. In the fourth [year], children could listen only to three classes, then they involuntarily put their heads on [their] desks.

More than a decade and a half later, the connections made by particular individuals were often explicitly indeterminate. For example, a woman in Selo made this observation: "We had a year once when almost every other day there was a funeral. We must have buried more than fifty people that year. Is it related to radiation? Who knows." (At least one Belarusian scientist I spoke to has reached a similar conclusion. According to him, these periods illustrate the effects of radiation exposure on more sensitive groups, such as people with chronic diseases.) Or consider the connections that the mathematics teacher in Selo made between radiation and children's memory problems: "I've been working here for thirty-three years, and I have not seen it as bad before." (She did not clarify why this cohort was affected but not the earlier cohorts of children who were exposed to higher doses shortly after the accident.) The teacher then generalized more broadly, expressing an opinion that I later heard from some Belarusian experts: "I think radiation simply makes every disease chronic."

Attempts to overcome this indeterminacy and uncertainty rhetorically, using, for example, the Soviet category of "practically healthy," are particularly interesting. Thus some teachers in Selo observed that "there are no completely healthy children in school." They referred to the prevalence of chronic diseases, heart conditions, gastritis, memory problems, and chronic fatigue. The teachers no longer attempted to draw precise connections between specific illnesses and radiation but instead made a broader argument about the effect of radiation on the local population. Their argument was based on health system practices and assumptions from the old Soviet Union: that a normal population has a significant percentage of children without chronic diseases—that is, "practically healthy" children. Even though the conditions that the children have vary, the observation that "practically healthy" children are an exception—almost all local children have chronic conditions—is then used to illustrate the effects of radiation. In other statements, lay commentators appealed to the uniqueness of health problems in the affected areas. For example, a member of an NGO working in highly contaminated areas maintained, "Where else do you see seven- or eight-year-olds with pacemakers?"

Although, overall, local residents did not doubt that increased levels of radiation affect one's health, the prevalence of particular local health problems and their connections to radiation exposure remained mostly unarticulated. Opportunities for such articulation, including those based on interaction with scientific experts, are extremely limited. The saleswomen at a local shop told me, "People get sick, but then they get some treatment (*podlechat'sa*) and they are okay." According to a group of local farmers, "Yes, people get sick. Depends on what they get sick with. They have flu.... People don't have 'thyroids' [i.e., thyroid problems], no. Don't know anybody who has a 'thyroid'." Asked if the problem of radioactive contamination of his village had touched him personally, one policeman (*milicioner*) replied, "No, I'm still young. But it still somehow shows up later. People are getting sick. Maybe it's because they are old, or maybe it's because of radiation."

Almost everybody, however, mentioned the impossible situation of being sick in a rural area, including the inadequate health-care infrastructure, the lack of qualified personnel, the long distances to hospitals, the long waits in overcrowded outpatient clinics, and the expectation that one has to bring "gifts" (i.e., bribes) to be seen by a doctor. These challenges to obtaining access to adequate health care further prevent any articulation of the relationship between health problems and past exposure to radiation.

Conclusion

Populations affected by invisible hazards cannot be assumed to be the most risk-conscious or to necessarily have special knowledge of the risks they are exposed to. Those who live with increased levels of imperceptible risks still have to learn to "experience" them—that is, to articulate the signs of radiation danger and radiation-related health effects. This work depends on available instrumental resources (tools and spaces for articulation) as well as interactive resources (opportunities for interaction with other perspectives).

Limited opportunities for articulation result in limited lay recognition of radioactive contamination and radiation health effects. Some instrumental limitations are obviously consequential, such as when there is no easy access to testing equipment or to reliable and clear results of testing done by experts. The lack of available public forums conducive to articulation on the communal level, including practices that could contribute to a popular epidemiology, is similarly consequential. Even adversarial forums such as Chernobyl-related lawsuits could have provided dynamic opportunities for articulation. Grassroots environmental organizations or other related civic movements are also largely missing in Belarus (notwithstanding a short-term rise of Chernobyl-related NGOs in the mid-1990s). A lack of broader and more adequate opportunities for articulation encourages laypeople to rely on administrative discourse (which, as we will see in chapter 3, is also strongly promoted by the state-controlled media).

It matters what kinds of articulation opportunities are available. Local residents are far more engaged in interactions with local and state administrative authorities (whose discourse is reproduced by the state-controlled media) than with scientific expertise. This administratively constituted signification of radiation hazards, when combined with a lack of other instrumental and interactive opportunities, can result in an underarticulated view of the hazard.

Belarusian psychologist Leonid Pergamenshchik commented that during his work in the Gomel region in the mid-1990s, he rarely head the word *radiation* used, except as a broad explanation for when somebody got sick. A decade later, the residents I talked to were clearly far more knowledgeable and articulate. But their articulations also reflected the administrative contexts that shaped them. More explicitly science-based contexts, such as radiological testing, narrowed the scope of valid concerns and interpretations, but it was precisely this organized and narrowed focus that made radiation risks more easily observable. Easier access to monitoring equipment, such as WBCs, and more engaged interactions with scientific experts—or, preferably, collaborations with them—would allow the affected communities to develop more elaborate articulations of the signs of radiation risks and effects.

Lay articulations are, in principle, not independent from other forums for articulation. Issues affecting scientific and administrative knowledge production, described in chapters 4–6, are also likely to affect the production of lay knowledges. (For example, Belrad has mobile equipment to test for Cesium-137, yet it only recently acquired equipment to test for radioactive Strontium-90, which might be a problem in some areas of the Gomel region.) At the same time, different population groups are likely to make different use of the available articulation opportunities, as will be illustrated in the next chapter. Different groups have different conditions for relating to the dangers of radiation and different motivations for learning about it. The result of these different approaches, in the words of Ulrich Beck, is that "someone who knows more and different things also sees more, sees differently, and sees different things."[18]

2   The Work of Living with It

More than two decades after the accident, the paradoxical fact about Chernobyl radiation is that individuals are responsible for their own internal contamination doses. The state food infrastructures have entrance and exit radiation monitoring, but people create their internal accumulation of radionuclides by consuming contaminated food from forests and private garden plots. The official media sometimes argue that people in the affected territories "have gotten used to radiation" because one cannot live in fear all the time (see chapter 3). At the same time, local perspectives on radiation danger are remarkably diverse, and individual contamination doses vary greatly. Who does what about radiation danger, and why? Why do some groups show such indifference to radiological advice and overall disregard for lowering their doses?[1]

This chapter will show that individual risk behaviors depend on a number of local structural factors. In other words, people make their own doses, but not in circumstances of their own choosing; these circumstances present a unique intertwining of radiological, geographic, economic, cultural, infrastructural, and other factors. For instance, forests, which can be described as a natural resource, have become integrated in the system of socioeconomic factors that results in greater internal contamination for the most economically disadvantaged groups. What appears on the surface as a problem of individual attitudes is, at least in part, a systematic, structural problem that blends together "natural" (radiological and geographic) and socioeconomic factors.

Furthermore, asking laypeople to make individual lifestyle changes—possibly with the help of public educational programs that make radiation danger visible—cannot address the underlying structural impossibility of this situation. Because radiological contamination is an ongoing problem, individuals and families have to exert significant, incessant efforts to mitigate the danger and reduce their own doses.[2] Mitigating radiation danger thus takes a lot of hard work—and that motivates at least some local groups to resist any interventions that make radiation danger more publicly visible. Many others worry about radiation only from time to time, when the problem is brought to their attention. Achieving a significant reduction of individual doses requires *infrastructural* solutions—solutions that can be sustained over time.[3]

To illustrate the work of living with radiation, this chapter explores a range of examples of risk behavior among the affected populations, including those who pay attention to radiation protection, those who appear to do nothing, and those who make intermittent radiation protection efforts. A puzzling, but not uncommon, attitude addressed at the end of the chapter is that individuals can be concerned with radiation health effects yet be indifferent to taking radiation protection measures.

This chapter is based on interviews with radiologists, residents of three affected districts, and representatives of other groups of the affected populations (see the appendix). It relies particularly on interviews with Minsk-based radiologists from Belrad, who have been traveling frequently to all the affected districts since 1996. Interviews with local residents—including a part-time radiologist in one of the villages in the Stolin district since 1991—were conducted during my trips to affected districts with Belrad radiologists and members of the CORE program.[4]

A Multiplicity of Perspectives

International reports on Chernobyl often describe the affected populations as anxious about living with increased levels of radiation (to the point of experiencing psychosomatic reactions), fatalistic about the effects of radiation, and overwhelmed by socioeconomic problems and poverty (see chapter 5). Apparently related to this fatalism is the fact that "only a negligible portion of households test food for concentration of radionuclides."[5] People are described as more concerned with socioeconomic problems than with radiation. A study by the NGO Belarusian Committee "Children of Chernobyl" argues that the affected populations are aware of the contamination of their food but people do not do anything to improve the situation; they often do not know any techniques for growing ecologically clean produce. According to the study, some of the residents claim there is nothing one can do "against radiation," so it is useless to try to take any measures. Others say it is very costly, and they lack the means. A popular quote from this study describes people living in the affected territories as thinking, "I would rather die from radiation than from hunger."[6]

These and similar reports (see chapter 5) aggregate different groups into a single "affected population" category. The affected groups are interpreted as sharing the same perspective on radiation danger and displaying the same risk behaviors. Their views are presented as irrational, intuitive, and experiential, and they appear not to change over time (see chapter 5). Yet there is no reason to assume one epistemologically homogeneous "affected population."

In this way, communities affected by Chernobyl fallout are similar to other "contaminated communities."[7] Martha Fowlkes and Patricia Miller, in their description of Love Canal and its residents' perceptions of the hazard, note that the community did not exist as such before the hazard; it was shaped by the situation, within the officially defined territorial boundaries. The residents also had to develop their own coherent interpretations of the new, invisible danger. The community was divided: some residents came to be minimalists in their definition of danger, whereas others were maximalists. According to the authors, these views corresponded to individual life-cycle factors and occupations. Minimalists were predominantly in or near retirement and had no children living at home; some were employed at local chemical facilities. All parents of young children, on the other hand, were maximalists.[8] Demographic factors alone do not determine one's beliefs, but they do put people in a particular position from which to seek information and evaluate evidence; as a result, some of the minimalists were intentionally uninformed. The minimalist and maximalist perspectives are "simultaneously and equally rational," however, in terms of an individual's interests and situation.[9]

Thus, depending on their circumstances, people may arrive at different perspectives on radiation danger. Perhaps even more important, one individual can hold multiple perspectives. I use the concept of perspectives to refer to narrative interpretations—of what constitutes radiation danger and where and when it might be found—offered in particular local dialogues at particular moments of their unfolding.[10] Since these perspectives are shaped within public dialogues, they are not, strictly speaking, individual perspectives. Individual opinions simultaneously echo established social perspectives and express interpretations unique to the person.[11] Furthermore, a person might be familiar with a number of different perspectives and can often provide meaningful accounts for and against a particular proposition (e.g., that the current level of radiation danger is unacceptable).[12]

In other words, most individuals can express different perspectives, and they can—and do—change what they argue depending on the context and with whom they are talking. For example, the same person might argue that

Chernobyl has had "grave health consequences" in some contexts (e.g., claiming Chernobyl benefits, teaching children, or talking to local administrators) while assuming a position of indifference in his or her daily life.

Although a person can hold many different perspectives, only one perspective can be acted upon at a time (in this sense, practical activity is monological).[13] In other words, one perspective becomes dominant at a given time in a particular practical activity—such as when an individual considers whether to move away from a contaminated area and prepares for relocation. This person is simultaneously aware of other, competing perspectives, which might become salient later, under different circumstances. Although only one course of action can be taken at any given moment, the individual can potentially go back and forth between competing perspectives, such as by following radiological advice and then not following it at a later time.

The radiation doses of people living in affected areas vary dramatically; their perspectives on radiation danger range from indifference to concern.[14] What, then, affects people's attitudes and behaviors? And what leads to particularly high doses in some groups? The next section explores how individuals accumulate their internal doses, which local experts consider especially dangerous.

Radioactive Forests and the System of Socioeconomic Privilege

The range of radiation doses in the affected territories reflects the intertwining of socioeconomic factors and factors that could be considered "natural." Local geographical factors (types of soil, prevalence of forests or fields) work together with radiological factors (patterns of radiation fallout, migration, and accumulation of radionuclides, or the comprehensiveness of initial decontamination measures) and socioeconomic factors (the size of the community, its socioeconomic traditions, and the well-being of its residents). The result is that socially and economically vulnerable groups are also the groups exposed to higher radiation doses.

The affected regions are mostly agricultural and face serious socioeconomic challenges. Most enterprises are unprofitable, and the regions are subsidized by the state budget. With some notable exceptions, personal business ownership and commercial farming are not developed. Local residents blame the lack of effective state policies, an inadequate legal system, and unstable state economic conditions.[15] People get by with low wages, private garden plots are maintained to supplement family diets, and in some areas produce is sold for additional income.

Difficult economic conditions lead to a greater dependence on free resources, including private garden plots, forests, and wild pastures. At the same time, post-Chernobyl radiological contamination tends to accumulate in the top layer of a forest or a field, and a number of practices effectively transfer radionuclides into private plots. For example, residents send their cattle to graze in wild pastures, then they consume radionuclides with milk and other dairy products—unless they have access to cultivated pastures purposefully planted with special kinds of grass that do not accumulate radionuclides. Contaminated cattle dung is used as a fertilizer, which contributes to the contamination of the private plots. Many local residents use forest wood in their furnaces, which turns them into "private reactors." The ashes are then used as soil fertilizer, continuing the cycle.

People also consume radionuclides in forest mushrooms and berries. Gathering wild mushrooms and berries is a popular leisure activity throughout the country, but it acquires special significance in the rural areas, where mushrooms and berries, along with fish and wild game, form a major free supplement to the family diet. Furthermore, privately grown or gathered food is sold at local markets and sometimes transported for sale in other parts of the country.[16]

Unlike individually grown produce, what is sold in grocery stores has to pass radiation control.[17] On the surface, then, individuals' internal radiation doses are the result of their own practices and lack of concern about radiation risks, since food production infrastructures have adequate radiation control. Yet as mentioned above, individuals create their own doses not in circumstances of their own choosing. Consumption of free goods from the forest might be less of a want and more of a need. Greater economic vulnerability translates into greater exposure to radiation risks, and this relationship is mediated, paradoxically, by the natural resource of forests.

The relationship among socioeconomic privilege, the use of forests, and risk distribution can be observed on the levels of both communities and individuals. For instance, based on the results of WBC testing, the residents of smaller communities tend to have doses of internal radiation two to five times higher than the residents of district centers (towns).[18] The explanation is simple: many smaller villages are worse off socioeconomically, and their residents rely more heavily on private plots and forests. (Larger towns also benefited from more comprehensive decontamination measures after the accident.) In the Soviet era, the government prohibited residents in the most contaminated areas from collecting food in the forest and consuming produce grown on family plots; these prohibitions were widely ignored. Such prohibitions were particularly ineffective in small villages, especially those located farther away from local centers.[19] There are simply fewer infrastructural and economic resources available in smaller, remote villages to replace private farming and the use of forests.

The relationship between community size and internal radiation dose was observed in three out of four districts studied by Anatoli Skryabin. One district, Narovlya, appeared to be an exception, and it is infamous for the high doses of its residents in communities large and small. The district is immediately adjacent to the zone of exclusion around the Chernobyl nuclear plant. According to Belrad radiologists, other contaminated districts typically have two or three "difficult" villages with particularly high internal doses, but even in the most difficult districts there are villages that are relatively "well-off." In the Narovlya district, all of the villages are considered difficult. In any village there, people have high levels of internal contamination, as measured by the WBC. Out of 3,000 children living in the district, more than 95 percent have been diagnosed with more than one chronic disease. Belrad radiologists relate these high internal radiation doses and the increased morbidity to the character of radiological contamination in Narovlya and the abundance of forests there.[20]

Aside from Narovlya, there are significant differences among other districts and among the communities within each district. Each community is marked by a different interplay of radiological, geographic, socioeconomic, and infrastructural factors. For example, Olshany village is located in the "difficult" Stolin district. This is one of the most densely populated districts, which results in a deficit of cultivated pastures and, consequently, higher levels of milk contamination. Olshany, however, has been less radioactively contaminated and has unique cultural and socioeconomic traditions that have resulted in significantly lower radiation doses. The village is sometimes referred to as the "cucumber capital of Belarus." Its economic well-being is easily observable. Bigger, modern-looking cottages in Olshany, with numerous cucumber "hotbeds" in between them, stand in great contrast to smaller, often unkempt houses in neighboring communities. According to Belrad radiologists, who have measured about 900 children there, "the numbers are all under 30 [becquerels per kilogram]"—which could be considered safe even according to Belrad's strict thresholds—and "there are significantly fewer problems." Islands of strong entrepreneurial (farming) traditions appear in several other locations in the Dribin and Vetka districts.[21]

The town of Bragin could be a counterexample. Located in an area with high levels of contamination (15–40 curies per square kilometer), the town has been largely repopulated during the 1990s. Many of the most educated town residents, such as doctors and teachers, left after the accident. Over the years, however, new residents arrived, including some refugees from other former Soviet republics who were fleeing unstable, even wartime conditions in their homelands. The town's reputation as difficult in terms of its radiological and socioeconomic conditions persists in the first decade of the twenty-first century. Another example of vulnerable communities is former residents of the 30-kilometer (18.6-mile) zone around the plant who were evacuated in 1986, often to other contaminated areas. Many have been living in houses built hastily; few have jobs.[22]

Individual doses within the same community can also vary greatly. Cases of particularly high or low doses show the influence of socioeconomic, occupational, and educational factors. For example, Skryabin measured internal radiation exposure with a WBC and then divided his subjects into four groups depending on the magnitude of their doses.[23] What he called the small-dose group was dominated by women (80 percent), and what he called the high-dose group was dominated by men (75 percent).[24] The small-dose group consisted mostly of white-collar workers and housewives, whereas the high-dose group was predominantly manual workers and retirees.[25] According to the head of Belrad's WBC Laboratory, children with the highest doses often come from poor or "at-risk" families:

We noted a long time ago that whether a child has significant accumulations [of radionuclides], how he feels, how he behaves—it depends on the social and economic status of the family. First, we were only noting it. Then we started asking school psychologists [social'nye pedagogi]. We approached them before doing the measurements in schools or while doing the measurements. A school psychologist has lists: children from poor [maloobespechennye] and at-risk [neblagopoluchnye] families. So, for example, in village Y in the Chechersk district, we make tests there eight times a year, and the top ten is always the same. Year after year, it's the same children who have the "top numbers."[26]

Other cases from Belrad's experience further illustrate that high internal doses are often tied to the same group of factors: the use of forests as a food source in combination with low socioeconomic status (see the examples in box 2.1). Children whose diet included food gathered in forests had the highest doses, as in example 1. Example 2 describes a family with no private plot, so its dependence on free forest goods is particularly significant, and the father of the family has been especially oblivious to radiological advice. The children in these examples had accumulations of thousands of becquerels per kilogram.[27] Belrad's recommendations suggest that internal accumulation should not exceed 200 becquerels per kilogram, with the control level established at 70 becquerels per kilogram.

Box 2.1

Episodes Illustrating the Role of the Forests against the Backdrop of Socioeconomic Factors

---

Transcriptions of interviews with Belrad radiologists

1. Forest Use in the Absence of a Private Plot

Village X in the Bragin district. Three children from the same family. They don't have a [private] plot. The children are always in the top ten [highest radiation doses among children in their school]. There are two other brothers from another family that deals vodka. They also don't have a plot. So since there is no plot in the family, and this particularly applies to teenagers [who are in] the seventh, eighth, ninth, and tenth years of school, the children want to eat, so they go gather mushrooms, or they go fishing, or they go with hunters. They basically eat whatever they can find. And their number is correspondingly large: four thousand becquerels per kilogram. There are similar stories in the Bragin district and practically everywhere else.

2. A Hunter's Children Have the Highest Doses

The highest doses we found were in village Y in the Narovlya district in 2000. The family has five children; their father is a hunter. Drinking vodka and hunting are all he does, instead of planting potatoes. Even in those conditions, it is possible to grow rather clean produce on treated soil; you can do it. But the man goes hunting, drinks a hundred gram [a shot of vodka] there, brings game home, and is even probably proud of it. He does not understand that what he feeds his children is poisoning them. The children had seven and a half thousand becquerels per kilogram.

3. A Doctor with the Highest Dose in the Village

Village Z in the Elski district. The [case of the] highest accumulation there was a doctor, the chief physician of the local hospital. You would think that if somebody knows about the [radiation] danger and how to avoid it, it would be him. It turned out that he loved hunting. Naturally, after you get a trophy [i.e., shoot an animal], you want to eat some of it. At least he had enough sense not to give it to his children. The children did not have high accumulations. He just stopped caring about himself, even though he was in his thirties, a young man.... He does not feel [the consequences] *yet*.

4. A Woman Who Loved the Forest but Had a Small Dose

A woman, headmaster of a school in the Chechersk district. She loved the forest—so much so that she even wrote in her will that she wished to be buried in the forest. So when she sat down in the chair [the WBC], all the teachers came running—now they were going to see some large numbers. But she only had fifteen becquerels per kilogram. Everybody was asking, 'How come?' She said that she only *walked* in forests; she had not picked up a single berry or a single mushroom since the late 1980s.

According to my interviews with the head of Belrad's WBC Laboratory, other children from the same villages could have much lower accumulations, of 15, 20, or 25 becquerels per kilogram—and parents who "pay attention." When Belrad radiologists conduct WBC tests in local schools, including kindergartens, the children are often too small to understand the radiation protection advice themselves, but a mother would "wait for the tests, then ask about the results, ask if they are high or not [and] what is the best thing to do. She would find it all out, and she would follow our recommendations and advice. She does it, and then we see the outcome." The WBC tests of these children typically show lower numbers. Belrad radiologists suggested that there are "not a lot of people like that, but there *are* people who do it." These are often better-educated residents, the local "intelligentsia." According to the head of Belrad's WBC Laboratory, "Besides knowing what to do, they also want to do it—they want to [raise] healthy kids. So once they find out, they are trying to do it." The contrasting illustration 3, however, demonstrates that there are examples of educated, well-to-do people with high doses.

Example 4 points to the importance of past radiological advice and how much it has been internalized by the local residents. The radiological contamination of local forests, the danger of eating mushrooms, and similar issues have become almost common sense. Whether these laypeople's conceptions are interpreted as scientifically correct depends on what expert perspective sets the criteria. The head of Belrad's WBC Laboratory argued that internal exposure accounts for 90 to 95 percent of the overall dose and that it is more dangerous than external exposure: "With external exposure, you come there and you get exposed, but then you leave and the exposure stops. If you get radioactive substances with food, they are inside the organism and are constantly irradiating from inside, which is more dangerous."[28] Regarding the woman in example 4, the head of Belrad's WBC Laboratory argued that the background radiation in her forests was "not that high. Up to 50 microroentgens per hour. It's not that much." The woman herself clearly knew not to ingest forest mushrooms and berries.

Individuals' interpretations of radiation danger, their understanding of radiological advice, and socioeconomic factors are intertwined in even more complex ways in examples 5 and 6 (box 2.2). Episode 5 shows an astonishing disregard for radiological advice in the face of economic benefits from so-called health recuperation trips abroad, whereas episode 6 provides a contrasting example in which radiological and socioeconomic concerns are not in conflict.[29]

Box 2.2

Episodes Illustrating Socioeconomic Factors

Transcriptions of interviews with Belrad radiologists

5. Consciously Feeding Contaminated Food to Children

This is a horrible case. I am not going to name the village. We were selecting children with large accumulations to go to Ireland for rehabilitation. In one village, they learned about it through their local radiation control center and started bringing produce to get tested. Usually, they did it rather laxly, but now they started bringing produce quite a bit. They were consciously giving "dirty" products to their children [so that the children will qualify to go to Ireland by having large doses]. They think that a month or three weeks in Ireland is a solution to their problems. The children will be happy and healthy. But this is far from the case. The harm that they have caused their children before the trip will not be compensated by the trip itself. We had one case like that. It ended up with a big scandal. The local authorities got involved.

6. Living on Aid Responsibly

Another approach: One village has a family with nine children. The windows in the house are taped instead of glass windows. The head of the collective farm [kolhoz] tells the man, the owner of the house, "I'll give you glass; put it in." But he does not want it because the Germans are going to come [bringing humanitarian aid], and they are going to ask who lives there. Oh, that's a family with nine children. They are poor [too poor to even have glass windows]; they need help. The Germans bring money, presents. The presents are then sold the same day near the local store. But when that man brought his children for [WBC] tests ... I have to give him that—he does not spend everything on his drinking. It was winter, and the children were dressed properly; you can't say that they had no warm clothes. So he has enough conscience. The children did not have high doses.... But he has an image of a sufferer, of a father of nine children. He does not work anywhere. He lives off the presents. Everybody knows that family now, and they try to pass a gift or something when there is an occasion, send money, products.

The examples mark the ends of the spectrum, and socioeconomic and demographic factors are particularly noticeable in these more extreme cases. The next section explores laypeople's perspectives on radiation risks from a different angle. Radiation is a continuous problem, and questions arise regarding the work of following radiation advice over time.

The Work of Living with the Imperceptible Hazard

Skryabin's study described above implicitly presupposes that the radiation-related behavior of the majority, those who fall in between the extremes of the highest and lowest doses, follows the same logic as that of the extreme cases, only to a lesser extent. A different perspective on radiation safety behavior of the in-between majority emerged from my interview with Aglaya, a local radiologist from the Stolin district.

The local center for radiation protection, where Aglaya was the only radiologist (working part-time; her primary occupation is as a nurse) was established by Belrad in the early 1990s, but it later became part of the state infrastructure.[30] The main goal of the center has been to test local food, and Aglaya's community turned out to have produce and wood with especially high levels of radiation. Faced with levels higher than any reasonable expectation, Aglaya single-handedly managed to draw the attention of international Chernobyl projects to her unusually difficult village, located in what was otherwise a mildly contaminated area.[31]

When asked whether she thought that people have just stopped caring and that "nobody does anything," Aglaya refuted that idea as follows:

I would not say that people just don't care anymore. I've been doing it [radiation control and dealing with people around these issues] for a long time now, since 1991. I think there are three groups of people. The first group is educated people who listen to what scientists and doctors say, what is said in the newspapers. The second group of people—and this is the majority of people—become worried about radiation from time to time, when there is something that brings it to their attention; for example, when there is a foreign delegation in the area dealing with Chernobyl issues, or if there is something in the news.... The third group of people is less educated; they just don't care, no matter what. The first and the third group of people are rarer; the majority of people are in the second group. People who bring their foodstuffs to be tested are the first category, and the rest ... they do it from time to time, when there is an occasion—when they get woken up, some test teams come, or foreigners come.

Aglaya's description of the second group as the majority that cares from time to time *when something makes the problem visible* was indirectly confirmed by the Belrad staff. Belrad had a number of local radiation protection centers such as Aglaya's in other contaminated villages. Radiologists in those locations were required each month to perform a specific number of tests on local food to check for radioactive contamination, but they could also do more tests, if necessary. Belrad reported that it received more test data from the local centers immediately after the Belrad radiologists had conducted WBC testing with their mobile units. For a certain period, the residents would bring more food to be tested, but then the wave would die out.

Similarly, when international projects organized radiation protection workshops or educational programs in the community—such as courses for expectant or new mothers—many wanted to participate. But then, according to Aglaya, "You have to make an effort and devote time to it, so many dropped out. Only the most persevering stayed." She added, "It is much easier to work with them [those who persevered], and they ask questions [and want to know]."

Radiation protection activities—such as getting produce tested, boiling meat in salty water for hours, or processing milk with a skimming centrifuge (because radionuclides remain in the whey)—are work. Living in the contaminated territories means that this work must be constant. At the same time, local residents often have only their own private, radioactively contaminated meat or milk. Aglaya described the complicated, almost pointless nature of food radiation control from the perspective of the local residents: "[People] are going to continue living here and continue eating these products. Nobody is going to bring them clean milk or berries. They won't be able to survive on the produce they get from a shop. They will continue having produce from their own plots." Aglaya acknowledged that people did not bring food to be tested "when left to their own devices":

People don't bring foodstuffs to be tested themselves, not much. When I go around collecting produce, they cooperate. But some are skeptical: "Are you going to give me money to buy new milk [if this is contaminated]? We have been eating it and will continue to eat it." But generally, I could always find an approach [to people], could always convince them and take measurements.

It appears then that many worry and pay attention to radiation danger for some time, when testing, educational workshops, or other interventions make it visible. These interventions produce a kind of fluctuating visibility of radiation danger. Yet individual radiation protection requires much work, and it is a challenge to manage and sustain such a routine in practice. It is not necessarily impossible, but it does require resources and motivation.

The following quotes from a social worker and then three farmers—who were responding to the question "What do you do when you know that your milk has too much radiation?"—illustrate this challenge and the reactions to it:

Nothing. Not throw it away, of course. I love mushrooms, too. And my grandchild is three years old. I want to give him something yummy. Milk, other stuff, we eat everything—we don't look.

Who would do that [all the radiation protection measures]? Farmers never have enough time.

We don't boil anything; we don't do this. We've gotta live here.

If you have money, then you can buy everything [i.e., food] clean. Apples are clean here.

A female nurse in her mid-30s said the following:

I'll tell you for everybody. Nobody boils meat [for as long as they are told to]. They just eat it as usual. Yes, radiation is something to worry about, but nobody does anything. At least I boil mushrooms when I marinate them; the rest are not doing even that.... I love mushrooms, myself. Am I supposed to not eat them? If only you could take a pill that would take radiation out—I would be interested in that.

A pill, in this case, would perhaps be the most effortless solution to the problem of radioactive contamination.[32] But making an effort to apply radiation protection techniques can also become a practical necessity. Artem, a CORE member, observed that people were concerned about producing clean produce not for themselves but because they had to have it tested before they could sell it. But he also left room for other attitudes: "I have not heard that people want to grow clean produce because that is what they want to eat. They are more concerned about it because they want to sell it.... As for themselves ... I have not heard it, but I suppose it depends on the person. Some might think about it, some might not."

These men and women who spoke of not testing their produce *were* aware of the danger, and in many cases, they did know what had to be done. Aglaya maintained that it is important to keep on testing and making the information available so that "they can at least choose the smallest of the two evils." According to her, "No matter whether people are willing to devote time to having their foodstuffs tested or not, it is important that even if they do it just once, they still learn, and they will know for the future." The situation has improved greatly since Aglaya first started working as a radiologist.[33] She noted the following:

People know now where contaminated places are. At the beginning, the first years [of her work, in the early 1990s] were particularly bad. People would bring milk, and it would have a couple of hundred or even a couple of thousand [becquerels per kilogram]; sometimes two thousand five hundred when the norm used to be one hundred and eleven and now is one hundred. I asked them where they brought it from, and the information spread that way.... The ones who do not know are the ones who never talked to me. There are people like that, too, who never once got anything tested. So I walk around the village myself [and] collect tests. And then [I] tell everybody; some listen to it.

Yet even simply collecting food samples for testing is not always easy. In Aglaya's experience, "People say that they know already. They are not cooperating very willingly. They know where radiation is around here." After an international project started working in the village in the 1990s, one more radiologist was hired to help Aglaya (for the rather symbolic pay of $10 a month) and found that "people sometimes tell you 'no' in a rude way." Aglaya added, "I have learned how to deal with them. I know who is doing what, who works where, who wants what. So with some I tell them that it's important for their children, with others, that they won't get hay."

Indeed, even with a greater awareness of which places and products were more radioactive, there were still people who picked berries everywhere. Aglaya reported: "Radiation fell out in spots here, so the bogs that are farther away are better. They fit within the [established contamination] limits. This is where I gather cranberries for myself and my family. But people pick up 'dirty' cranberries, too. Everything is picked up everywhere. Nothing is left behind."

This is not necessarily a sign of a lack of awareness. Aglaya has submitted a project proposal with an international organization to create a map of radiological contamination around the village based on the results of her tests of the produce.[34] When asked if she thought there would still be people who, after looking at the map, would gather mushrooms or berries in the most contaminated spots on the assumption that there would be less competition there, she replied that this was "exactly how it is going to be."

At the same time, Aglaya insisted on the importance of making radiological information available even when there was a lack of initiative from many local residents. Educational workshops and programs in her community often sought to attract young people, who were generally more willing to learn and implement changes.[35] "There is work with young people in the medical center, and at school, teachers talk about it, too," she explained. The local medical center (*ambulatoria*) had booklets describing, for example, how to reduce radiation in milk by making cream or cottage cheese from it. The problem, however, is that the educational efforts were tied to a relatively short-term international project. Aglaya stated the following:

When they leave, all of this will probably stop, so that's why I'm saying that it would be good to have these kinds of lessons at school on a permanent basis. Something like "how to protect yourself from radiation" or "how to live in these conditions"—I don't know what to call it. But just so people are not indifferent and that they are not neglecting these issues, since I still think that radiation does its ugly job.

The educational and information programs are often temporary, but the problem is ongoing. Aglaya's ideal solution would be to make the programs permanent—as part of the local infrastructure and educational system—"since we are living with this kind of problem." Children's education at school acquires particular significance: "If children had those lessons at school, they would teach their parents, and then they would know when they become mothers themselves. If only schools had those kinds of lessons, with their own teachers, with a set curriculum."

But even the most well-informed and willing individuals cannot solve an ongoing problem of this scope by themselves. Radiation protection requires taking administrative steps beyond just educating people and beyond encouraging them to develop what some international projects call "practical radiological culture."[36] Under the circumstances—when radiation exposure is affected by the system of socioeconomic privilege and when radiation protection requires ongoing effort—focusing exclusively on teaching laypeople to protect themselves is asking them to do the kind of work that is in principle beyond what they can do. Such a pervasive, ongoing problem demands infrastructural solutions. For example, many residents burn local wood in their furnaces, which produces ashes that are extremely radioactive. People have been advised not to use ashes as fertilizer and to bury them outside their gardens instead of just throwing them away. But as a local resident put it, "For all these years, imagine how many 'tombs' I'd have around here." The organized removal of ashes is a recognized need, yet it is seldom done.

Even making radiation danger visible in each community requires infrastructural solutions. Measuring radiation in food and humans cannot be done with cheap and portable Geiger counters. It requires expensive and maintenance-intensive equipment and, consequently, must be organized on the administrative level for an entire community or district.

Aglaya brought up administrative solutions when she talked about "over-the-top" contamination of food in the early 1990s and how people gradually learned. She explained what would happen when people brought radioactive milk to be tested:

I asked them where they got it from, and the information spread that way. I told them that they have to demand that their hay [be] changed to clean. Now the state is more cooperative. It is easier to get cultivated pastures, and there is less wild hay. But there are still shortages. If somebody does not work in the collective farm, they might not get it either. Or somebody who has many cows.... So it happens that people are still stocking up on wild hay.

Local administrations are often resistant to sustaining the visibility of radiation danger and to taking on additional financial burdens by creating infrastructural solutions.[37] Some have the attitude that since radiation itself is invisible, the problem is not the radiation but rather those who attempt to make it visible: "When there is no information, there is no problem," stated one local radiologist summarizing this attitude. Belrad radiologists met this kind of resistance in some district administrations when they attempted to check radiation doses in communities that had lost their status as radioactively contaminated and had been excluded from the list of affected localities.[38] Belrad staff contacted several district administrations asking to confirm their agreement to cooperate and assist in the organization of the tests. The Mozyr district administration was distinctly discouraging, as the head of Belrad's WBC Laboratory described:

We were told the following: People have made some noise on that topic, we've calmed them down, so don't disturb our people. Don't come here, we do not need this. Everything here is good, everything is clean, everything is calm. Why this reaction? If there are problems, it means that there should be benefits that have to be fought for. There has to be funding, and so on. And otherwise, when there is no information, there is no problem. And when there is no problem, you don't need resources for it.

A driver traveling with the Belrad radiologists commented privately that the team frequently had problems with local administrations, "especially with those who might be responsible in any direct way. They try to create obstacles for us in any way because 'everybody [in their care] is healthy.' One local doctor told us that we 'get in the way.' He said he had his own WBC, but we never saw it."

Some local interests would prefer that the problem remain invisible since it might create work, put strains on already limited resources, and make administrators personally responsible. Indeed, whenever Belrad radiologists conduct WBC tests in schools, they typically give the list of doses to the school officials so that they can "take the appropriate measures." In cases of particularly high doses, this could create problems for both school and district administrators.[39]

Aglaya brought up local community action, which she considered necessary since less and less was being said in the media about radiation. She said, "If we are not going to raise that question [of radiation] locally, it will soon die out completely." The head of Belrad's WBC Laboratory similarly emphasized individuals' initiative and persistence in dealing with local administrations, but he also observed that negative examples outnumbered positive ones (see box 2.3): "I would not say that everything gets fixed like this—there are more negative cases."

Creating and implementing infrastructural solutions is particularly difficult, since once they are enacted, these solutions acquire new meanings in the context of "other things happening," not just on the basis of their effectiveness in terms of radiation protection.[40] For example, one way of reducing radiation doses in children is to provide them with free school lunches. Schools get their food supply from state enterprises that have entry and exit radiation control; consequently, the lunches at school contain fewer radionuclides than the produce from private plots that the children eat at home.[41] When children stop getting free lunches at school, according to Belrad, "their WBC measurements go up."[42] At the same time, in rural areas, there is typically only one school for several neighboring villages, and only the children who are from the villages officially classified as contaminated receive free lunches. The distinction creates a social problem in which some students get lunches and others do not, even though the teachers try to remedy the situation and feed all the children, since there are often students who are sick or absent. Moreover, not all children residing in the officially contaminated territories are eligible for lunches. Children of refugees from other Soviet republics living in the Gomel region often lack residency papers (*propiska*) and thus cannot receive lunches, either. Further complicating matters, the teachers may be motivated by interests unrelated to radiation protection. From the teachers' perspective, without school-provided lunches, the students are more likely to skip the last class—they have "less of a tie to school," to quote a teacher from the Gomel region. Radiological issues are hardly ever the sole concern, and the effectiveness

Box 2.3

Examples of Local Action and Dealing with the Local Bureaucracy

---

Transcriptions of interviews with Belrad radiologists

7. A Rare Positive Example

In the Luninetsk district, there is village N; the school there has 450 children, not a large school. The principal of the school is very interested in solving these kinds of problems. One of the schoolteachers is also a radiologist at the radiation control center. They are both very concerned for the children and work to solve the problem. They achieved great success. They got the local authorities [*raispolkom, sel'sovet*] interested in their information. They started conducting workshops on the basis of their center—first a district workshop, then a regional one—addressing the questions regarding how to solve this problem. Then they asked the regional authorities to help with "clean" mixed fodder for the cattle and with some other issues. The issues actually got resolved.

But this is a rare case. When the principal called me, I told him to send me copies of all the documents because it is such a rare case, so that I can show other people that if there is enough motivation and enough confidence, this can be achieved. I want to show it as an example because many say, "So we write a letter to the authorities and then what? Nothing is going to come out of it." Of course, this kind of pessimism is justified to a large degree, but there are positive examples.

8. "Fixing" the School Stadium

In one of the villages, they asked us to measure background [radiation] on their school stadium. We did, and it turned out to be higher than acceptable limits. The school reported to the regional authorities [*raiispolkom*]. The regional authorities sent the sanitary team [*komanda sanstancii*] to take measures. So here are the measures they took: the sanitary team made the tests and gave a prescription—prohibit physical training lessons in the local school stadium. Now they all had to be conducted inside. The situation is quite laughable. The sport lessons are inside, but during the breaks, everybody runs outside and hangs out in the stadium. And after school, that's where everybody hangs out, too. But "measures have been taken."

It's a question of expenses. The battle went for a year and a half. We, from our side, also asked and made demands. Eventually, it really was done. The trucks came, took off fifty centimeters [nineteen and a half inches] of the top layer of the soil, brought clean soil, planted grass, and that was it. Of course it cost money, but these are gigantic results. The stadium is practically "clean" now.

Box 2.3 (continued)

---

9. Cultivated Pastures

In the Gomel region, one of the local radiation protection centers has a device for measuring radiation in food. And they suddenly detected "dirty" milk. The head of the district administration asked us to come with the chair [the WBC]. The local children had accumulations of Cesium-137 higher than one hundred [becquerels per kilogram], which is fairly atypical for that place. It turned out that the local collective farm was given money by either Comchernobyl [the State Committee on Chernobyl] or the regional authorities, I don't know, to cultivate pastures for the local cattle. The local collective farm got the money and spent it, and the locals did not get a peat bog for cattle pasture, but rather they got the dirtiest place in the neighborhood. [The administration] probably knew it was "dirty." ... With some joint efforts of the village authorities [selsovet] [and] the local doctor, the head of the hospital, the money was found, the residents got their pasture, and milk became clean again. We came with the chair again, and the accumulations got better.

---

of infrastructural solutions depends on more than how well they work to reduce radiation doses.

Mitigating radiation danger is work, and people who live with radiation are faced with a job that might exceed their resources—especially since people with fewer resources tend to be the ones who are exposed to greater radiation risks. Mitigating these risks requires infrastructural solutions. But whether these solutions are effective in terms of radiological protection is hardly ever the sole concern in implementing and assessing these infrastructures; it is just one part of an indivisible web of other concerns and circumstances.

Visibility and Temporality

The previous sections show that the idea of the "affected population" must be disassembled into a multitude of situated perspectives. But the groups described above are in the metaphorical bull's-eye of the present scope of contamination. The affected population is not limited to the residents of the areas officially recognized as contaminated; many other groups might be considered affected populations with a varying degree of certainty. Among them are Chernobyl cleanup workers, evacuees, and resettlers from heavily contaminated areas, communities that have lost their "affected" status, and many other Belarusian—or, broader, European—residents exposed to radiological fallout and contaminated food soon after the accident. Some of these groups lost their "affected" status and related benefits in recent years (see chapter 3); others were never officially recognized as such in the first place. What is worth noting is how the spatial recognition of contaminated communities implies a particular temporal relationship to radiation risk (recognized as still present) and Chernobyl as an event.

For many, though not all, outside of the areas currently recognized as affected, radiation exposure is something that happened in the past—it is a matter of *having been* exposed. The effects trail behind the exposure—they might be in the past, in the present, or expected to reveal themselves in the future. When people move between areas labeled contaminated and "clean," they might also be reinterpreting the temporality of their radiation risk. Defining the temporal dimensions of radiation risk—what is in the past and what is still there—is as ethically charged as defining the spatial boundaries of contamination. The more the danger is located in the past, the less need there is to pay attention to radiation safety in the present. Reframing Chernobyl as a thing of the past might be another strategy of rendering it invisible (as we will see in chapter 3 in some examples of official discourse).

When I returned from Aglaya's community to Minsk, my trip and the motivation behind my research was challenged by another interviewee, Galina, who questioned the research agendas of scientists and social scientists who continued to worry about the existing dangers. Galina explicitly prioritized justice and care for those who were already sick over preventative radiation protection measures:

We are worried that people boil their meat for three hours, but that's nonsense. Don't take it personally, but nobody has ever done it and nobody will do it. But what can be done, nobody wants to do.... Everybody knows that society can be considered respectable only when it can take care of its own old people and disabled.... It seems to me no other country has as many disabled as we do. Tell me, are you worried about the disabled that we have already? ... Today, eighteen years after the accident, I don't think there will be more disabled people because meat won't be boiled for three hours.... I think people die *indirectly* from Chernobyl. When they see the hopelessness of the situation, they start drinking, smoking—they forget about their health. Nobody is going to labor with meat here, and it is important that correct moral priorities are set. And maybe then, they'd even boil meat.... I am not against your research. I am surprised when people are fighting for what still can be, or what can potentially happen, and not what has already happened.

For Galina, concerns with present-day radiation protection are not realistic—she used the word *utopian*—because the main harm has been caused already. What society should do is "honor and take care of its disabled"— that is, those who have already been affected by Chernobyl radiation. She deplored the lack of attention to the health care of the Chernobyl cleanup workers and their loss of benefits and monetary compensation.

Most of the time, Galina was certain that we already know who has suffered from radiation and which health effects are radiation-related. She referred to the disabled who have the officially established "Chernobyl link" that qualifies them for benefits under article 18 of the law On the Social Protection of Citizens Affected by the Catastrophe at the Chernobyl Nuclear Power Plant.[43] Galina was more skeptical when talking about the health status of the Chernobyl cleanup workers, whose benefits are also described in the law but are not explicitly tied to the workers' health. "I cannot tell you," she stated. "I'm not a doctor, but then a doctor would not be able to tell you, either, what [the health data is] and how it is collected, because there is a chance that it's just not [collected]."

Galina's own deeply situated interpretation—she and her husband participated in the cleanup, and she now resides in Minsk—both draws on and questions official categories. She raised ethical questions that complicate what should be made visible: Is it ethically right to care about present and future risks while the consequences of past exposure go unrecognized? Ethical visions are grounded in particularly situated viewpoints, but in this case, the viewpoints are also marked by divergent temporal relationships to radiation danger.

Five personal stories in box 2.4 illustrate this. It is notable that none of the five speakers (who work with international Chernobyl projects or local Chernobyl-related NGOs) are indifferent to the problems of Chernobyl. These stories illustrate the complexity of the temporal perspectives and ethical agendas surrounding the question of whether we should still be concerned about radiation danger. Their perspectives are informed by their (re)locations but also, especially in the cases of Aglaya (narrative description 1) and Tonya (narrative description 2), by the gendered work of caring for family members in the face of unknown and imperceptible dangers.

Box 2.4
Narrative Descriptions of Personal Chernobyl Experiences

---

1. Aglaya (Woman in Her Late 40s): Has Already Experienced the Effects, but the Danger Continues

Several days after the accident, Aglaya and her fellow villagers were told that they should prepare for resettling. "People were planting potatoes then, they got back from the field black with [radioactive] peat." In the end, the village was not relocated, but the summer after the accident, all the local children were sent to Moldova. Aglaya's husband, a teacher, went with the children, including their six-year-old son. The sight of their departure was "horrible," and "the picture of the village after everybody has left was horrible."

There wasn't much information after the accident, but there were rumors. Some people (mostly administrators and doctors) took their families and left. Some people left later, in the early 1990s. Aglaya's village was in the zone of voluntary resettlement. Many returned later—"either their jobs did not work out, or maybe their home was calling them back." Aglaya pleaded with her husband to leave, but he did not want to because of a concern that both of them would not find jobs, that they would not find good housing, or that people in a new place would not like them. Aglaya tried to have more children but had two miscarriages. "At the time, many women had similar problems.... And then it calmed down. Maybe because the iodine hit [*iodny udar*: radioactive iodine decomposed soon after the accident] passed by, or maybe the bodies adjusted."

Aglaya was one of the first local radiologists trained by Belrad. When she started working as a radiologist at the local center for radiation protection, some of the tests she sent to Belrad were returned for retesting because the numbers were too high. Aglaya tells matter-of-factly that ashes from her own furnace saturated the device's detection capability, which topped out at 37,000 becquerels. By repeatedly dividing the amount of ashes in half until the radioactivity read less than the saturation limit, Aglaya calculated the radioactivity of her ashes to be 200,000 becquerels per kilogram.

When the Belrad experts did not believe her, Aglaya brought the ashes with her to Minsk to demonstrate their radioactivity. She was concerned that other people were using similarly radioactive ashes as fertilizer for their plots. A relative, a TV journalist, had Aglaya appear in a program on radiation. After that, people started coming to her village, including later "the foreigners," members of international projects. Aglaya's son was sent to work (*raspredelenie*) in village Y in the Mogilev region, which is also in the affected territories. She insisted that radiation "should not be forgotten about" and noted that it is not a good sign that "they still cannot find a doctor to work here. They even built two houses, but nobody wants to come here."

Box 2.4 (continued)

2. Tonya (Woman in Her 40s): Has Experienced the Effects; in the Contaminated Areas, the Danger Continues

Tonya, head of a Chernobyl public foundation, was in her early 30s and lived in Minsk when the accident happened. Her daughter was two and a half years old. Months later, Tonya got pregnant. "Nobody told me anything. I really think doctors should have been telling women to wait with having children." Her son was born in September 1987, and when he was two and a half, he was diagnosed with leukemia. The daughter also had health problems.

Tonya wrote letters to everybody, "even to Seattle," trying to save her son. More than a decade later, she was still thinking about what she could have done that would have saved him. Doctors told Tonya that nobody proved his illness was from radiation, but she pointed out that in the beginning, the doctors were also saying that thyroid pathologies were "isolated cases" and "look how it all came out." According to Tonya, "Parents of those children do not have a doubt, even if they are not well-educated. Mothers become as knowledgeable as doctors. But that only happens when the misfortune strikes."

Protecting her family from radiation was still important to Tonya. She would never make soup with bones (which accumulate strontium) or buy Stolin cucumbers. Nobody in her family has been allowed to gather or buy mushrooms, and she did not trust the official story on TV but instead read independent newspapers "to not be fooled like an idiot."

"If I have an option," she said, "and life, of course, makes its own corrections, I try to be cautious. After all, I have a daughter, and my daughter has to have children. As the saying goes, 'The man is not going to cross himself until it thunders.' Radiation danger exists only for those who faced the fact of a child's illness, and maybe also for people with enough education to behave themselves differently."

According to Tonya, "People should be informed. Some would say that it is rubbish, others would behave differently. But nobody should go through what I have gone through. And what about those cats? Cats with two tails? Other deformities? Where does that come from? There have been some mutations."

Tonya provided anecdotal evidence of Chernobyl's effects based on the experiences of her friends who are doctors and her personal experience of interacting with cancer patients. She was convinced that the number of cancer patients increased after the accident.[44]

3. Victor (Man in His Late 30s): Radiation-Related Experiences Confined to the Past; No Clear Danger Now

When the accident happened, Victor, now a local coordinator for an international Chernobyl project, was a teenager living in the Stolin district. Back

Box 2.4 (continued)

then, the accident "did not affect his life much," although his village was in the zone of voluntary resettlement. He went to college in Minsk, and after graduation returned to the Stolin district and stayed in village M. He worked at the local collective farm, but like many in village M, lived off the cranberry trade. Wild cranberries were gathered at nearby bogs (known to accumulate radiation), and sold in Moscow.

The problem, however, appeared to be not the business but rather the conditions in the house in which he lived. They were so bad that Victor developed nephritis. While in the hospital, he was tested for internal radiation using the WBC. The results were so high that "they sent people to the village and tested everything there; radiation in mushrooms was four hundred times higher than the norm." When asked why he ate those mushrooms, Victor said, "The people whose house it was would cook something and put it on the table, and everybody sat down and ate. Nobody was looking [at what to eat]."

The doctors told Victor that he was young and had to get away from there, so he decided to leave the village. He still lives in the same district, and not too far from village M. For Victor, the Stolin district is just "where I live." When asked about radiation in the place he lives now, he reacted rather dismissively: "It's in a good spot."

4. Aleksey (Man in His Late 20s): No Clear Signs of Danger in the Past or Present; the Effects Have Already Happened

Aleksey, a manager for an international Chernobyl project, was a young boy when the accident happened. His family lived in the moderately contaminated areas, but it did not affect him much; in his words, his life was "parallel to it all." Nobody had any fear about radiation. Nobody, including his parents or teachers, told him that anything was dangerous. The only possible connections to Chernobyl were annual health examinations at school, but nobody paid much attention to them, either.

"When it was most dangerous," he said, "there was no information about it at all. And then people got used to it, they started to forget. People only remember about radiation when they start having problems with health. Absolutely nothing ever reminded me about Chernobyl." According to Aleksey, the only thing that reminded him of Chernobyl now was his job. He never thought that trips to the contaminated areas could affect his health. Nobody who hired him ever asked whether he had concerns. This was never a question.

The only thing about Chernobyl that Aleksey related to and talked about with more interest was the question of Chernobyl-related compensation and inconsistencies in the Chernobyl laws. Aleksey was upset that his current Minsk residency disqualified him from particular benefits. "What difference

Box 2.4 (continued)

does it make if I live there, or I live in Minsk? I have already been affected, haven't I? And if I live in the Minsk region, it does not make it easier for me." He does not know whether Chernobyl has had any effects on his health.

5.   Galina (Woman in her 50s): The Danger and Effects Are in the Past

Galina, the head of a Chernobyl-related NGO, and her husband lived in the town of Narovlya. Her husband was in the military and took part in the accident cleanup. He became sick and was hospitalized several months after the accident. He left the hospital "a disabled person" and later "received article 18," official recognition that his disability was tied to the accident.

Galina herself had the status of a "liquidator" (cleanup worker, defined by article 19); she was helping with the evacuation of children to summer camps after the accident. When asked if she was concerned about radiation, she replied with stories of her experience arranging these evacuations in May 1986, and she could not help crying, "The town without children is a scary place." Galina decided not to let her own two daughters be sent to these summer camps with the rest of the town children, but instead she arranged for them to stay with family in Riga. In December 1986, after her husband had been decommissioned, they joined their daughters in Riga and returned to Belarus only in 1993 (they stayed in Minsk instead of coming back to Narovlya).

Galina herself has had health concerns. Although she and her family stayed in Narovlya for only several days after the accident, one of the daughters has thyroid problems. Galina's mother died from cancer, and the house was impossible to sell because "nobody wanted to buy a house with radiation." Galina was very upset with the lack of attention to the health care of "the disabled of Chernobyl" and the cleanup workers, but eighteen years after the accident, she believed that the radiation danger was negligible.

Conclusion

The seamless web of radiological, geographic, and socioeconomic factors leads to the disproportionately higher exposure of the most economically vulnerable groups. As we have seen in this chapter, internal radiation doses are not distributed evenly. The affected population is not a homogeneous group. Different groups and different individuals hold a variety of different perspectives, shaped in the context of their particular life circumstances. Vulnerable rural populations are likely to rely on subsistence farming and a heavier use of forest goods, which remains the key dose-contributing factor.

Relatively high and relatively low individual radiation doses illustrate the role of socioeconomic factors especially well. Yet socioeconomic characteristics do not explain everything. We have seen that people's attention to radiation protection and what people do in practice may fluctuate, at least in part because of the changing public visibility of local radiological contamination.

Residents of the contaminated areas have generally become more educated about radiation protection in the years since the accident; their exposures are, in general, smaller than in the first years after the accident. Yet Belrad radiologists claimed that the doses are not decreasing on their own; progress depends instead on local authorities and their efforts. The challenge faced by local communities is infrastructural in nature: radiological contamination is a pervasive and ongoing problem, and sustaining constant radiation protection efforts demands incessant work. In the absence of adequate state and local infrastructures, the burden of this work falls on local residents, who might not have the resources or motivation to engage in it or sustain it. Meanwhile, the paradox of infrastructural solutions, such as providing free school lunches, is that they are not likely to be implemented or assessed solely on the basis of their radiological effectiveness.

Finally, the affected populations extend beyond the still officially contaminated territories. Including these broader communities lets us consider significant differences in how various groups and individuals understand the spatial and temporal boundaries of radiation danger—and the extent to which drawing these boundaries depends on personal circumstances and viewpoints. We could describe these differences as variations in interpretation of "space-time work." For example, to see the area as no longer significantly contaminated is to redefine the radiation risk as in the past, thus canceling out the need for continued radiation protection work. We have seen these differences in personal accounts. The next chapter describes how this boundary was being redrawn in the official discourse in the media.

3   Waves of Chernobyl Invisibility

In a May 1986 essay entitled "Anthropological Shock," Ulrich Beck reflected on the aftermath of Chernobyl as a "media event" in Germany. By *anthropological shock*, Beck was referring to the experience of the inadequacy of our senses when faced with radiation danger—human senses register nothing when exposed to increased levels of radiation. Individuals' own sovereign judgment is rendered impossible. Without the information provided by the media and other social institutions, laypeople would not even notice the increased levels of radiation. Beck describes this as "the experience of cultural blinding." Another aspect of the shock is that "those who until now have pretended to know don't know either. None of us—not even the experts—are experts when it comes to the atomic danger."[1]

We need not assume that people remain "cultural[ly] blind" to radiation danger—that no cultural ways of identifying danger are developed. Nevertheless, the two previous chapters have shown that neither can we assume that those most affected will necessarily develop ways of "seeing" the danger and become the most risk-conscious. As argued in the introduction, one's experience of radiological contamination, even for those who live with it on a daily basis, is highly mediated—by media narratives, authoritative accounts, scientific theories and equipment, maps, rumors, and any other representations that render radiation publicly visible. In Belarus, media portrayals might be one of the most influential means for representing post-Chernobyl radiation, especially since these portrayals are dominated by the official discourse, which in turn is echoed by practical changes in the lives of the affected populations, such as loss of monetary compensation and other benefits (see chapter 1). Almost everybody I interviewed in Belarus on the topic of Chernobyl could clearly describe the official position on the matter.

As one of the main forces in making radiation publicly visible, mass media can raise awareness in local contexts (especially since there is no perceptible difference between the contaminated and uncontaminated regions). Mass media can also potentially set a national agenda on radiological contamination.[2] But the media can also help displace imperceptible risks as a matter of public discussion. Media representations of Chernobyl and radiation danger shift with time and changing political agendas; they do not become increasingly more accurate and comprehensive. The scope of Chernobyl's consequences can be enlarged or diminished; the nature of these consequences can be—and was—redefined and reframed.

This chapter follows the metamorphosis in how Chernobyl's consequences were represented in the Belarusian media—their growing and shrinking in scope. I invite the reader to trace what the media were referring to when they spoke of Chernobyl: What areas are contaminated, how badly are they contaminated, and how long is this contamination going to last? Are the consequences already present, or what consequences should be expected, and when? What are the solutions? A young Belarusian woman's comment to me in Minsk in 2005 summarized the more recent outcomes of Chernobyl's media transformation: "Chernobyl's consequences have been shrinking for years now. Old statements about Chernobyl being an 'international problem' sound funny. The affected area in Belarus has shrunk and seems to be roughly equivalent to the size of an airport." The puzzle this chapter considers, then, is how the effects of Chernobyl first grew and then shrank, and why.

The disappearance of Chernobyl from the official, government-controlled media cannot be explained by changes in the actual scope of contamination. It would also be ambiguous and even misleading to describe this disappearance as the result of cultural forgetting. Chernobyl did not simply disappear; it was extinguished in waves as a result of particular types of framing that set the health effects of the fallout outside the scope of immediate concern. Before that, the greatest public salience of Chernobyl did not begin until about three years after the accident, a period of great political transformation only two years before the collapse of the Soviet Union. This eruption of visibility was followed by the gradual transformation of the Chernobyl discourse that again paralleled broader political transformations in the country. I trace the metamorphosis of Chernobyl as a historically situated, politically determined discursive construction.

The answer to how and why Chernobyl disappeared as a radiological problem relates to the broader issue of what kinds of representation can potentially promote and sustain public attention to environmental risks, especially imperceptible risks with delayed health effects. There are at least two challenges to sustaining media and public attention to these kinds of risks. First, making the risks observable requires theories, devices, and methods of science (see the introduction).[3] Yet science makes for a difficult topic of mass media reporting. Second, science itself is also a site of conflict over meanings and agendas.[4] Among other things, economic motivations, including those of state governments and industries, work to limit the extent to which risks are acknowledged. This chapter specifically addresses the place of economic versus radiation safety concerns in the media coverage of Chernobyl. Given the importance of scientific apparatus and vocabulary for making radiation risks visible, we will also consider the role of the scientists.

Extensive government control of the media in Belarus means that the discourse has been dominated by the official perspective. Government-controlled media have offered the most comprehensive framework for interpreting the consequences of Chernobyl. This framework has been presented relatively consistently across government-controlled television channels, radio, and newspapers. I refer to this perspective as the *government discourse*. Oppositional newspapers started appearing in Belarus in the mid-1990s. As we will see later, risk-related interpretations in the oppositional media have been constructed as a reaction to the position of the official media and thus are in part determined by it. Specifically, the reframing and disappearance of the consequences of Chernobyl in the official discourse in the mid- to late 1990s was met with symbolically overloaded, dramatic, and even hyperbolic representations in the oppositional media. I describe this phenomenon as the *hypervisibility* of Chernobyl.

The analysis that follows relies on a comprehensive sample of Chernobyl-related articles in the government-controlled and widely read *Sovetskaya Byelorussiaya* (*SB*) (later renamed *Belarus Segodnya*, but still abbreviated *SB* and often referred to by the old title). In addition, I consider coverage in three other newspapers: *Gomel'skaya Pravda* (*GP*), a local, also government-controlled newspaper from the most heavily affected region; *Ekologicheski Vestnik* (*EV*), a regional ecologically focused newspaper; and *Narodnaya Volya* (*NV*), a national oppositional newspaper established in the mid-1990s. Since the media coverage of Chernobyl has never been entirely coherent and devoid of contradictions, I emphasize the differences among periods rather than provide an exhaustive description of any one time. My descriptions of the dominant themes are based on the content analysis, and particular examples are drawn from the discourse analysis of each period.

The volume of Chernobyl-related coverage in *SB* and *GP* peaked in 1989–1991, the last years of the Soviet Union, fell sharply in 1992, and then showed a rather cyclical character, with coverage dominated by anniversaries of the accident. (Figure 3.1 shows the increases in 1996 and 2001.) As the volume of related coverage generally declined, Chernobyl increasingly became "what we talk about in April."[5] But the peak of 1989–1991 stands out from this pattern. We now turn to what happened during that period, as well as how these transformations were both built upon and undone later.

The Eruption of Visibility and the Role of Scientists

At the time of the Chernobyl accident in 1986, the Communist Party's ideological control remained pervasive. Much of the early coverage of Chernobyl in *SB* appeared directed at containing the perceived scope of the accident in both space and time. Reports described the "liquidation of the accident," implying that the accident was about to be fixed and eliminated in its entirety. Military cleanup workers were referred to as "liquidators." The coverage appeared under the rubric "The Zone of Special Attention" (*Zona*

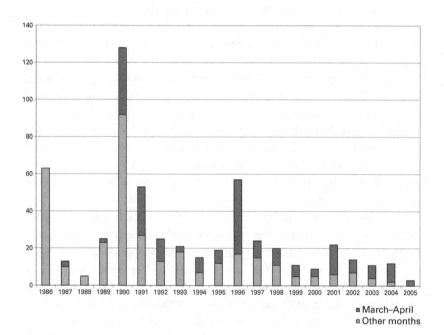

■ March–April
□ Other months

Figure 3.1
Articles about the Chernobyl accident and its consequences (including articles published in March–April versus other months), *SB*, 1986–2005.

osobogo vnimaniya), designating specifically the 10 kilometers (6.2 miles), or later the 30 kilometers (18.6 miles), around the Chernobyl nuclear plant. This was the "zone of *temporary* evacuation"; it was *"temporarily* closed." No mention was made of the radioactive fallout that was blanketing most of the country. Instead, *SB* quoted the Nobel Prize–winning physicist Carlo Rubbia as allegedly saying that the catastrophe "had been prevented," although it also included more cautionary voices suggesting that even though there was no danger, "it is better to be safe than sorry." Later, the focus expanded somewhat from "liquidation of the accident" to "liquidation of the consequences of the accident," thus admitting there were consequences to handle.[6] By omission and juxtaposition, the reports suggested that those territories that were not immediately adjacent to the Chernobyl nuclear power plant were not contaminated.

GP offered similar containment strategies, soon placing most of the Chernobyl coverage under essentially the same rubric "The Zone of Special Attention" (*Zona asablivyh klopatau*). Only a few, positive articles on Chernobyl appeared in the summer: People evacuated from the "zone" were welcome in new places, and children (evacuated to summer camps) were enjoying their "happy vacations." Agricultural production had not changed; it was business as usual.

Experts contributed to this exercise of discursive containment. Very few people outside Moscow's official party and scientific circles knew the extent of the consequences in Belarus; the Belarusian response to the accident was coordinated from Moscow.[7] In interviews and reports from press conferences in May 1986, Soviet and international experts buttressed the position of the Soviet party leadership.

These top scientists included Leonid Ilyin, the director of the Institute of Biophysics of the Academy of Sciences, USSR; Hans Blix, the director general of the IAEA; and Robert Gale, an American leukemia expert from the University of California, Los Angeles School of Medicine, who flew to Moscow to treat the highly exposed first liquidators. According to these and other experts whose opinions appeared in the press at the time, there was no risk for the general population; the levels of radiation were not dangerous, and they were diminishing.[8] In late 1986 and in 1987, top scientists quoted in the articles emphasized the continuing importance of nuclear power. Only two articles featured scientists who were neither recognizably part of the Soviet scientific establishment nor international experts supporting the position of that establishment. Only one article used a local Belarusian scientist as its source.

For the next year and as late as 1989, *SB* journalists continued to echo scientists' claims that there was no danger because the doses were low. In *GP*, an extensive 1988 article claimed that scientists had found that not a single person had a confirmed disease related to radiation exposure—the main problem was said to be radiophobia, fear of radiation.[9] Yet that article also represented an early sign of change: while repeating opinions expressed at a scientific conference on Chernobyl, the report also criticized "radiosecrecy," the bureaucratic withholding of information about the consequences of Chernobyl, and "radioindifference," the lack of reinforcement of adequate radiological control. It was not only the criticism but also the sheer size of the article (almost a whole page long), that made it stand out from the earlier coverage.

The first significant shift in how the media defined the consequences of Chernobyl followed the 1989 resolution of the Communist leadership to develop the five-year Program for Overcoming the Consequences of the Catastrophe at the Chernobyl Nuclear Power Plant (hereafter referred to as the Chernobyl Program). The top-down decision to consolidate the efforts at mitigating the consequences of Chernobyl—efforts previously coordinated by numerous isolated decrees—allowed for a discursive shift that, given the political climate of the final years of the Soviet Union, led to a rupture. The very nature of the task changed the temporal perspective; the consequences of Chernobyl could be, and even had to be, approached not as a *temporary* but as a *long-term* problem.

The growing visibility of Chernobyl was made possible by the climate of political transitions and greater political openness and the emergence of an identifiable political opposition, the Belarusian National Front (BNF). Even before the official establishment of the BNF in June 1989, the oppositional intelligentsia was the key force pushing towards the 1989 declassification of the documents related to Chernobyl and calling attention to the massive contamination of the republic.[10] These efforts translated into disagreement with the Soviet experts.

One of the first publicly observable episodes of this controversy appeared on February 9, 1989 in a report published in *SB*. It was unusual in several respects, not least because it was a report from a public meeting of the Communist Party's Commission on the Liquidation of the Consequences of the Chernobyl Accident (held on February 2). The Soviet administrators were advocating the same old answers, yet at the question-and-answer session, journalists asked new questions. They challenged radiation control procedures, asked about the health effects of radiation, and raised the question of additional evacuations: "When will people be able to buy individual dosimeters?" "Many doctors left the zone—isn't that a sign of radiological danger there?" "If there are no adverse health effects, why are people receiving double wages plus 30 rubles compensation for each family member living in the zone of radiation control?"[11]

Years later, Tamara Belookaya, the head of the NGO Belarusian Committee "Children of Chernobyl," told me that the oppositional intelligentsia were actively working with reporters on those questions.[12] The same February 9 issue of *SB* included the first public maps showing broader radiological contamination in the republic. The publication of increasingly large and more detailed maps followed.[13]

SB continued to print reassuring interviews with experts of the Moscow school, including Professor Ilyin, in the spring of 1989.[14] The precise claims these experts were making, along with the Belarusian scientists' objections to them, will be discussed in the next chapter. For now, suffice it to say that the disagreement revolved around the *concept of radiation protection*, meant to serve as a yardstick for the design and implementation of the Chernobyl Program. The concept of radiation protection referred to the official, science-based assessment of the post-Chernobyl radiation risks and of the necessary protections. The Soviet scientists proposed "The Concept of Safe Living in Areas Contaminated after the Chernobyl Accident." In this assessment, Ilyin and his colleagues did not consider the post-Chernobyl exposures to pose significant health risks. According to them, no additional resettlement or other radioprotective measures were necessary. These arguments were echoed in an interview with the French radiation safety expert Pierre Pellerin entitled "No Secrets about the Nuclear Power Plant." The French expert painted radiation-related concerns as unreasonable: "Nobody in France has asked for individual dosimeters."[15] From the perspective of the Soviet experts and international experts like Pellerin, people—meaning local, Belarusian people—suffered from "radiophobia," and their fear was not based on objective, scientific information.

Belarusian media audiences first saw a coherent opposing perspective, that of the local scientists, in the summer of 1989. The Supreme Council of the Byelorussian Soviet Socialist Republic (BSSR) was scheduled to discuss the Chernobyl Program. It was a sign of the changing times that the coverage preceding this session included the opinions of Belarusian experts who explicitly disagreed with the Moscow-based scientific authorities and their Safe Living Concept. Inclusion of both viewpoints, not just the official party line, was justified as information for both the general public and the deputies about to debate the Chernobyl Program.

The Belarusian scientists proposed their own, alternative concept. From their perspective, people could not live where it was not possible to obtain uncontaminated food and where normal life activities had to be limited. Journalists writing for *SB* appeared rather sympathetic to this outlook, presenting it as more sensitive to the actual context of living in the affected territories. The Belarusian scientists thus not only presented some opposition to Moscow-based experts; even more important, they instigated the discussion of the stakes behind the Soviet concept, uncovering the politics behind what was earlier presented as technical and objective.

This debate continued during the Supreme Council session. *SB*'s transcripts of the session included references to Belarusian, Moscow-based, and international experts. The session officially gave Chernobyl the status of a national disaster. Some deputies pressed for a more unified and scientifically based concept of radiation protection, emphasizing that this would help raise funds for mitigating the consequences of Chernobyl. When the Chernobyl Program was finally published in October 1989, it adopted the views of the Belarusian scientists that people should not live where they could neither produce nor be supplied with uncontaminated food products.[16] This meant additional evacuations from the most affected areas. Economic and administrative challenges of providing accommodation for the evacuees became a topic for discussion in the government-run media. At the same time, *SB*'s journalists grappled with the implications of the new approaches to Chernobyl, even suggesting, "It is time to stop dividing the zone into more contaminated and less contaminated parts. Radionuclides, even in small amounts, do not add to people's health.... We will have to leave—sooner or later—all of the sick land."[17]

The period of greater visibility of Chernobyl followed the 1989 Supreme Council session and the publication of the Chernobyl Program. *SB* regularly published maps of contamination and tables describing background radiation. Official accounts now acknowledged that almost a quarter of the republic's territory was contaminated. By the end of 1989, the discourse around Chernobyl unambiguously represented it as a long-term problem. The extent and the quality of the coverage began to change. *SB* published 100 articles on Chernobyl in the first half of 1990 alone, compared to 81 in the previous three and a half years. The 1990 session of the Supreme Council of the BSSR, which focused on the adoption of Chernobyl laws, was televised in full, and *SB* published its complete transcripts, pages and pages of speeches and discussion from the session.[18]

Greater public recognition of radioactive contamination meant more discussion of Chernobyl's enormous socioeconomic costs and its administrative complexity. The 1990 session contended with both sets of challenges.

The debates were particularly heated, for example, on compensation for those living and laboring in the contaminated territories.[19]

The controversy around a scientific concept for radiation protection that would inform all decision making was not completely resolved. Some deputies were still troubled by diverging expert opinions and bemoaned the lack of "scientific consistency" in defining a "coherent scientific concept of safe living on the contaminated territories." Nevertheless, the Supreme Council adopted the concept proposed by the Belarusian scientists as the foundation for the Chernobyl Program and drafting of the Chernobyl laws; Ilyin's Safe Living Concept was described as "immoral."[20]

Top officials, meanwhile, emphasized economic constraints. Deputy Prime Minister Aleksandr Kichkailo argued that "the scope of the activities [mitigating the consequences of the accident] is completely determined by our material resources." Prime Minister Vyacheslav Kebich sought to remind the deputies that "elimination of the [Chernobyl] consequences and [transition to a] market economy are tied together." The country's material resources, however, were extremely limited. The last years of the Soviet Union, according to one *SB* journalist, were a time of shortages "of almost anything, practically down to bus tickets."[21]

With a greater acknowledgment of the staggering scope of Chernobyl came wider political struggles. Protests and strikes occurred in Minsk and Gomel. *SB* acknowledged the turmoil in the summer and fall of 1990, but it did not cover all the strikes; some earlier protests had been organized by the BNF in 1989.[22]

Strikes in Gomel protested the absence of "real action" by the government. The Chernobyl Strike Committee in Gomel even organized a march to Moscow, called the March for Survival, to meet with the Communist Party authorities after having witnessed "how helpless the Supreme Council was" in dealing with Chernobyl.[23]

GP included more reporting on local protests, even publishing the resolution of the strike committee in Gomel.[24] The resolution called for, among other things, a law protecting the rights of citizens affected by the accident, for individual dosimeters, and for a ban on temporary (raised) radiation norms and thresholds. Ironically, a Soviet-style report praising the Chernobyl efforts of the local administration appeared right next to a report of the strike.

The problem of a lack of resources for dealing with the consequences of Chernobyl became increasingly salient, especially given inadequate assistance from the Soviet government.[25] As V. Yakovenko, the president of the socioecological union "Chernobyl," put it in *SB*, "Before there was not enough openness ..., now there are not enough resources."[26] Indeed, the articles discussing socioeconomic issues far outnumbered those referring to issues of radiation safety during this period of heightened visibility (see figure 3.2).

SB started publishing calls from civic foundations that were fundraising for the recuperation of the "children of Chernobyl."[27] Calls for assistance and cooperation were also directed internationally. These appeals, as well as some other articles of that period, were loaded with increasingly strong, emotionally charged expressions referring to the accident: *catastrophe, national disaster, tragedy, calamity, the wound of Chernobyl,* and *our sorrow.* Chernobyl-related coverage in *SB* became far more emotional in 1990 than it was in 1986, when only a few articles had used such epithets; practically none did in 1987–1989. Most of the epithets referred to the accident and its aftermath, very few referred to radiation and radiation danger, and none that referred to radiation were used widely or consistently.

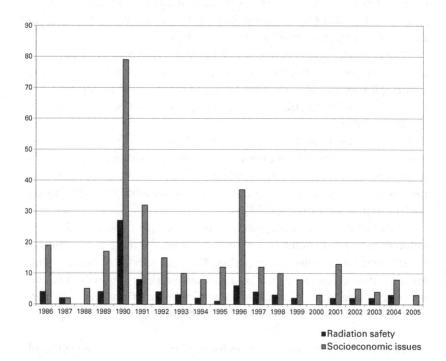

■ Radiation safety
■ Socioeconomic issues

Figure 3.2
Chernobyl-related articles discussing radiation safety issues and socioeconomic issues, *SB*, 1986–2005.

By 1991, numerous stories were describing the experience of living in the affected areas, including problems with radiation control and the production of uncontaminated food, along with the socioeconomic problems related to evacuation. An article entitled "Chernobyl: The Concept and Life" claimed that "more than 90 percent of people residing in the areas of strict ... radiation control consider relocation to be a necessary measure of their protection."[28] The Communist Party leadership in the republic interpreted Chernobyl-related social unrest in a broader political context, as the "struggles to change the sociopolitical regime." A. A. Malofeev, the first secretary of the Communist Party of Byelorussia, stated, "There is not enough funding for the Chernobyl Program, but that would only become possible with stabilization of the [broader political] situation in the country." At the same time, A. A. Grahovski, the first secretary of the Gomel committee of the Communist Party, blamed scientists and a lack of consistency in scientific recommendations for the earlier resettling of the residents of the most heavily affected communities into areas that were still significantly contaminated.[29]

In 1991, however, scientists—who had played the key role in exposing Chernobyl's massive consequences in the media—started gradually disappearing as sources of SB expertise on Chernobyl. The publication of radiological maps also became less frequent after 1991. The peak of scientists' presence in the media was before and immediately after the 1989 and 1990 sessions of the Supreme Council discussing the Chernobyl Program and Chernobyl laws. Some claimed that the Soviet Safe Living Concept had "destroyed public trust in science"—reflecting, perhaps, that this kind of explicit, public controversy among experts was not common in the Soviet and early post-Soviet contexts, and it might have challenged some public assumptions about science.[30]

SB still featured some experts as sources on Chernobyl-related issues other than radiation protection. These experts often had heavy titles and figured as particularly qualified politicians, explicitly pursuing political agendas, offering political recommendations, and lobbying for certain state programs (and not limiting themselves to the discussions related to their research). In the three years that followed (1992 through 1994), science and scientists as sources almost completely disappeared from Chernobyl-related coverage; their presence was limited to the discussion of the activities and research carried out by international organizations and experts.

The coverage in GP sheds some light on why scientists became gradually less relevant as sources on this issue. GP published extensive materials on radiation safety and even letters from readers checking, for example, for consistency among the reports or asking how to translate background activity levels into the density of radioactive contamination. In 1990, *GP* also started publishing *Ekologicheski Vestnik (EV)*, a monthly supplement filled with maps and tables describing contamination levels.

Much of the information on thresholds and norms, including their visual rendering as maps, was provided by government sources. Heads of related government agencies appeared as sources much more frequently than science experts did. The use of these sources came with no discussion of government policies and interests; their interpretation was not contextualized or questioned. In other words, outside the context of the Chernobyl Program and laws—and the related explicit disagreement between scientists of the Soviet school and their vocal Belarusian opponents—technical information again appeared objective. Providing such technical information was the domain of the government agencies conducting the measurements, and there appeared to be no need for independent experts to interpret or contextualize it.

Such an unwavering belief in the objectivity of technical information is particularly ironic in light of *EV*'s broader coverage of Chernobyl. *EV* frequently raised issues of secrecy and public trust in official sources in the context of stories with less technical data, such as stories on administrative violations of radiation safety procedures. A careful reading of *EV* also tells us about other limitations on the visibility of Chernobyl during that period. *EV* was one of the most critical publications, working hard to educate the local population about radiological risks. Yet despite the number of local stories published from explicitly critical perspectives, any type of overarching, systematic analysis was rare. Instead, incidences were presented as local, isolated cases.

The Economic Reframing of Chernobyl and Subsequent Normalization

The recognition of the massive scope of contamination in Belarus brought public attention to the costs of mitigating the consequences of Chernobyl. From the beginning, the Soviet government's public response had included compensation for those who continued to live and work in the heavily contaminated areas. Adoption of the Belarusian concept of radiation protection as the foundation for the Chernobyl Program and then two Chernobyl laws were about to increase Chernobyl-related costs, which already went beyond direct radioprotective measures. One of the two 1991 laws, On the Social Protection of Citizens Affected by the Catastrophe at the Chernobyl Nuclear Power Plant, established significant economic privileges and compensation for the affected groups, such as liquidators and those who were classified as disabled from Chernobyl, whose health problems had demonstrable connection to the accident. The benefits included, for example, free summer camps for children, free use of public transportation, admission to a university, housing or an interest-free loan to build housing, and even the right to privatize existing housing, in some cases.[31]

The question of how to pay for the Chernobyl Program was a concern even before the disintegration of the Soviet Union. Indeed, it was one of the arguments offered for why the republic should stay within the collapsing Soviet Union.[32] Without support from outside, it was reasoned, there might be no resources for mitigating the consequences of Chernobyl. But the dissolution of the Soviet Union at the end of 1991 left the republic alone in dealing with the consequences of the disaster and implementing its new Chernobyl Program and Chernobyl laws.

In the first years after independence, the republic faced growing deficits, inflation, joblessness, and poverty. This lack of resources made adequate financing of the Chernobyl Program impossible. In addition, the administration of the program was often corrupt and inefficient, and the national leadership was unprepared to deal with mitigating Chernobyl's consequences after the dissolution of the Soviet Union.[33] From organizing relocations and building houses for evacuees to organizing dosimetric control and responsible farming practices, the mitigation efforts were fraught with problems: money was spent on hopeless projects, appropriated, and misdirected.[34] The value of compensation for living in the highly contaminated zone had been eroded through inflation. As described by David Marples, some affected areas were "heading for the socioeconomic abyss."[35]

The challenges of mitigating the consequences of the accident were not just problems of poverty, compounded by massive political and economic transformation; the response to Chernobyl was made even more difficult by deeply entrenched Soviet bureaucratic practices. We should not forget that the Soviet Union failed to develop legal and bureaucratic procedures for recognizing and mitigating ecological crises throughout its history (which had already included Mayak nuclear disaster in the 1950s). Increased openness, or *glasnost*, in the last years of the Soviet Union could not compensate for the vast cultural, political, and infrastructural unpreparedness.

The combination of the 1991 laws and independence brought a marked shift in the official discourse on Chernobyl. Radiological dangers were no longer merely weighed against economic concerns; rather, economic concerns surfaced as the main issue. From 1991 to 1995, the government-controlled media gradually reframed Chernobyl as an *economic* catastrophe as much as a nuclear accident or a radiation safety disaster. Chernobyl-related coverage turned to socioeconomic issues, including the difficulty of securing funds for the implementation of the Chernobyl Program. Chernobyl was now referred to as a catastrophe because of the scope of the required mitigation effort, the lack of finances, and the inadequate state response. The economic framing tied the well-being of the affected areas to the economy of the state, emphasizing the price of the Chernobyl accident and the lack of material resources to solve Chernobyl-related problems.

From 1992 to 1994, the majority of Chernobyl-related articles in *SB* mentioned economic problems, and nearly a quarter of all the articles discussed international humanitarian assistance or other forms of economic assistance from abroad. Other stories called for local assistance. Fundraising appeals were made at all levels, from government bodies to some newly formed NGOs and even to private citizens.[36]

The Supreme Council issued an international appeal for assistance, especially for children's health recuperation. Appeals were also made to the United Nations. *SB* followed the activities of international organizations related to Chernobyl but also reported on and even promoted fundraising efforts within the country, such as the first national TV marathon in 1992.[37]

Newly formed Chernobyl foundations—quite a few of which used *children of Chernobyl* in their names—were also looking to raise funds. *SB* published direct appeals from such organizations but also produced critical reports with such titles as "Is Humanitarian Assistance Humane?" and "Where Did the Money Go?" The former report questioned the motivation of local humanitarian foundations. According to this article, these foundations sought to arrange children's recuperation abroad because they were "not spending their own money [but were rather spending donations]."[38]

The use of strong epithets and metaphors amplified the calls for assistance and the discussion of socioeconomic challenges. One could still find dramatic descriptions of the radiological aftermath: the "nuclear clock of Chernobyl," a "radioactive tornado," "nuclear ashes spread by the nuclear volcano," the "ashes of Chernobyl," "treacherous radiation," and "our bitter lot." Appeals for assistance also brought up health problems, which were assumed to be related to the disaster, as in the article titled "The Name of the Killer—Radiation."[39]

Similarly, a published fundraising letter from an NGO in the least affected area, the Grodno region, mentioned that 165 local children had recently been diagnosed with cancer and stated in its headline that "Child-Killer Chernobyl Is Continuing Its Black Mission."[40] Government officials and journalists both assumed the causal link between Chernobyl and health problems. According to the news story "Chernobyl Footprints," the minister of health observed the worsening health of the Belarusian population.[41] One might speculate that the visibility of Chernobyl as a radiological problem was still strong and that there was no need to justify or explain the connection.

Newspaper coverage continued to comment on humanitarian aid, especially assistance from foreign NGOs, into the mid-1990s.[42] These accounts suggested that the funds may have made a difference for many families. In a survey by a leading sociological agency reported on by *SB*, residents of the contaminated areas viewed the humanitarian aid as more efficient than assistance from the state; 57.6 percent in 1994 and 72.7 percent in 1996 expressed positive attitudes toward the international humanitarian assistance.[43] At the same time, *SB* and other news media offered few attempts at a systematic assessment of the effect of humanitarian aid. NGO assistance from abroad thus did not radically transform either the official discourse on Chernobyl or the challenges the state was facing in fulfilling the plan for mitigating the accident's consequences.

With the economic reframing, the public visibility of all aspects of Chernobyl began to decrease. The volume of related coverage in the state-controlled media decreased. The focus of the coverage—the kinds of Chernobyl-related problems being discussed and the solutions proposed—was also changing. From 1992 through 1994, much of the coverage dealt with evacuations, emphasizing their cost. After 1994, socioeconomic problems in the contaminated areas (unrelated to relocation) came to be the most frequent topic.[44] The focus of Chernobyl-related coverage narrowed to the most contaminated areas, and it was about to shrink further.

The first—and, to date, the only—Belarusian president, Alexandr Lukashenko, was elected in July 1994. Soon thereafter he began steadily consolidating power; in November 1996 he held a national referendum to expand presidential authority. *SB* reported that the president would now personally supervise the Chernobyl question. Toward that end, he began visiting the contaminated regions every April. The official coverage of Chernobyl refocused on administrative efforts: how the government was helping the affected populations, including the state government assistance to the contaminated areas, and what challenges local administrations faced. This discourse presented Chernobyl as a massive economic problem in need of administrative solutions. Reflecting this frame, *SB* journalists routinely asked, about a couple of months after Lukashenko's annual voyage, "What has changed after the president's visit?"[45]

Although discussions of Chernobyl still frequently invoked strong meta-phorical references, the words *catastrophe, disaster, tragedy,* and *wound* now referred more to post-Chernobyl decisions and their consequences than to the accident itself. In other words, in 1995–1997, *catastrophe* and similar descriptions applied to what happened *after* the accident. This was in part connected to expressions of skepticism about international assistance on the state level, such as "international society is not going to help us." Yet in 1995, Lukashenko still maintained that Chernobyl should remain an issue in UN programs, and he disagreed with the IAEA's claims about the absence of radiation-related morbidity after Chernobyl.[46]

In 1996, Lukashenko again appealed for more international assistance, reminding the international community that Belarus carried "the main burden of the consequences of a global accident." In September of that year, however, the president declared to the national audience that "the approaches are going to change."[47] Emphasizing that it was impossible to relocate everyone, the government—and the *SB* articles that trumpeted the government's view—argued that it was important to create normal socio-economic conditions in the contaminated territories. The new policies aimed to reestablish normal life conditions for residents of the affected, mostly rural areas.

By the time of Chernobyl's 10th anniversary, *SB*'s coverage had been showing early signs of the shift. The paper reported that some local authorities believed that scientists had miscalculated the "level of survivability of people in the zone" and that "it is not scary to be sick, one can get better." Descriptions of the contaminated areas reflected a qualitative change. A new kind of report depicted them as places where people wanted to stay and where they were not necessarily in grave danger: "It is dangerous to live here, but there is no desire to leave" or, in the words of a local retiree who refused to resettle, "Those who left, still die.... Those who stayed, death does not take." Indeed, the headline for the 1996 presidential visit to the affected areas emphasized that people were returning to their contaminated communities: "People, Like Birds, Are Returning to Native Nests."[48]

With an economic-administrative framework ascendant, the topic of Chernobyl's health and radiological consequences continued to erode. Increasingly, reports might discuss administrative or economic issues in highly contaminated areas without ever mentioning the accident or any radiological considerations. For example, the official discussion of what was called the rebirth of the village did not pay much attention to radiological contamination, effectively blurring the difference in reporting between the problems of the contaminated areas and those of other regions. Aside from the wave of coverage around the 1996 anniversary of the accident, significantly fewer articles explicitly referred to Chernobyl.

The geographic scope of the areas officially recognized as contaminated also shrank during this time. Some reports mentioned a new concept of living in the contaminated territories (which the critics compared to the earlier Soviet Safe Living Concept), the adoption of which would cancel benefits for some affected populations.[49] In April 1996, *GP* published a list of localities that no longer had the status of affected (which had brought with it some compensation for the residents and some limitations on agricultural production). *SB* also observed that there were fewer "hot spots on the map of the republic," but it did note that risks from internal exposure through the consumption of food produced in the contaminated areas remained.[50] The status changes were presented as reflecting natural processes; there was no indication of secrecy, ignorance, or miscalculation. When discussions of radiation occurred, the associated dangers were presented as significantly more tolerable than the earlier discussions had made them out to be.

The discourse of normalization only intensified after 1998. News reports about re-creating normal life in the contaminated territories continued to focus on the government's efforts at solving socioeconomic problems, although they also mentioned building infrastructures for radiation safety. The president continued his annual April visits, frequently choosing locations within the Gomel region, close to or even within the zone of exclusion, thus acknowledging only a limited scope of consequences. *GP*'s coverage of those visits added to Chernobyl's visibility, but only as a narrowly framed economic-administrative problem.

In 2001, the president began his visit at the Polesski State Radioecological Reserve, encompassing some of the most contaminated areas. There he spoke about agricultural production and rehabilitation: "This is our land. I believe that it will be in use again [*vostrebovana*]." The residents of an evacuated village in the exclusion zone who had occupied empty homes against radiologists' advice offered the president traditional welcome treats—bread with salt, milk, and honey—and he accepted the offers (the report does not tell if he actually ate them). The same image was repeated in 2004, when *SB* wrote, "Radiation hasn't left [these places] completely. But people learned to grow rather clean vegetables." Again, the president was said to gladly accept offers of local food.[51]

The president's annual visits—now the high point of the dwindling coverage of Chernobyl (as shown earlier in figure 3.2)—inevitably promoted a discussion of the remaining economic-administrative problems and the government's proposed solutions. To encourage economic development, areas with significant, but not the highest, contamination were given the status of free economic zones, which included tax benefits. Meanwhile, the question remained of what to do with semiabandoned villages that had lost the status of primary resettlement areas (where the residents were entitled to relocation) but where some residents still remained—leaving them in need of basic infrastructural support, such as access to groceries, health care, transportation, and telephone service.[52]

The persistent emphasis on socioeconomic issues implied that the problems were largely the result of *responses* to the accident, notably the spontaneous and organized relocations. Little mention was made of any adverse consequences of the radioactive fallout itself. According to the official government line in *SB*, people had calmed down and realized that "nothing irreversible has happened, and that, with some willingness, it is possible to return life to its usual course." The repeated sentiment was that "life overcame Chernobyl" or "life endures."[53]

Belarusian scientists, who could potentially review the findings of the post-Chernobyl research, remained nearly absent from this official coverage, despite claims that the rehabilitation policies implemented on the contaminated territories had "scientific grounding." From 2001 to 2005, only three articles referred to the perspectives of scientists, including an interview with Yakov Kenigsberg, the deputy director of the Institute of Radiation Medicine and Endocrinology, who acknowledged only thyroid cancer as related to post-Chernobyl radiation exposure. The official discourse defined radiation health effects narrowly, as only those related to the thyroid.[54]

With the passage of time, Chernobyl had also become a recognizable symbol, "even history."[55] Despite the fact that Chernobyl's boundaries were being redrawn at the time, the 10th anniversary coverage presented Chernobyl as something with identifiable shape; it stood for a recognizable set of issues. It was used, for example, as a synonym for *hell* or as a shorthand reference for any extremely tragic occurrence—within articles on subjects completely unrelated to the accident or its consequences. For example, an article describing the evils of alcohol dependency was entitled "Drinking Is Worse Than Chernobyl."[56]

A couple of years later, Chernobyl was not only assumed to have a clearly defined scope, it now also appeared as just a historic event—the nuclear power plant accident, which was exclusively in the past, severed from the subsequent remediation efforts or dangers that still lingered. Finally, a new type of story described Chernobyl in connection to socially nonproblematic events, such as the tale of a 116-year-old woman living in the contaminated

Bragin district, descriptions of excursions to the Chernobyl plant organized in Ukraine, and accounts of Cold War spies in the Chernobyl zone after the accident.[57] This new type of story addressed human-interest topics without addressing anything about either the Chernobyl accident or its consequences—nothing related to radiation safety, health, economic issues, government actions, or people's experiences in the contaminated areas. The strong administrative focus of the remaining coverage made it more difficult to interpret what problems, if any, were attributed to Chernobyl.

The discourse of a nonproblematic, historical Chernobyl was occasionally mixed with dramatic references and images. Indeed, these rare, dramatic, but often unsubstantiated references stood out against the dominant discourse of Chernobyl as an economic problem. For example, a story describing local young painters from an exhibition devoted to the Chernobyl accident interjected that "many [of the painters] have died already."[58] Similarly, a 1997 report on the construction of the Center for Radiation Medicine in Gomel included a picture of two bald, presumably sick children with the construction of an administrative building in the background. A 1995 report clamed, without providing evidence, that, "every year 10 or 11 Belarusian cleanup workers … pass away in their prime—under 40 years old."[59] Some of these isolated reports voiced sentiments reminiscent of the early 1990s, such as when a 1998 story claimed, "The Chernobyl disaster is not only real, but according to scientists, it will continue to remind us of itself for years to come."[60] These voices were rare in *SB*, but as we shall see, they occurred with more frequency, in *NV*, the oppositional newspaper.

The early years of the 21st century saw more of the familiar attitude ("And still life!" or "We will not give up our land") and the same discourse of normalization, now mixed in with praise of the government efforts.[61] These efforts were said to be unparalleled among the affected countries, and they included a system of radiological control, agricultural measures, pensions for the disabled of Chernobyl, tax benefits, credits, free lunches for schoolchildren, and nine health rehabilitation centers. The president marked the 25th anniversary of the accident by going to the most contaminated land, the Polessky Radiological-Ecological Reserve, and arguing for the development of tourism there as well as the agricultural production of apricots, grapes, and apples—which, according to scientists, absorb little radiation. The official discourse was emphasizing "sustainable development," not "rehabilitation." In 2013, Anatoly Rubinov, the chairman of the Council of the Republic, declared success in overcoming the consequences of the disaster.[62]

The most notable development of the first decade of the 2000s—and an indication of how engrained the discourse of normalization has become— was the government's decision to build a nuclear power plant in Belarus. The first references to these plans appeared as early as 1997, when *SB* asked, "N[uclear] P[ower] P[lant] in Belarus: Experts Approve. Who Is Against?" and offered a discussion of the economic benefits of nuclear power without any mention of radiation safety.[63] Fifteen years later, the decision to build a plant was becoming increasingly real. The president acknowledged resistance to the idea of building a nuclear power plant in Belarus but dismissed the concerns, asking, "What are we afraid of? There is nothing to be afraid of." From the perspective of the Belarusian government authorities, "Chernobyl should not stand in the way of modern energy production in Belarus."[64]

Starting in 1992, the Belarusian state-controlled media gradually reframed Chernobyl as primarily an economic-administrative problem. This framing, along with the later emphasis on "normalization," contributed to increasing public invisibility of the radioactive contamination and its potential effects. The spatial and temporal scope of recognized radiation risks shrunk; the volume of coverage decreased. There was now little notice of the radiological consequences of Chernobyl. This was as true for *GP* as it was for *SB*, although *GB* paid slightly more attention to the remembrance of the accident and the emerging "Chernobyl culture." Throughout the 1990s, *GP* also included a few more references to science, but these very infrequent stories did little to challenge the government discourse or bring attention to radiological issues.[65]

Similar to *SB*'s reporting, the discourse of normalization in *GP* worked by strategically appropriating the position of the people who stayed in the affected areas, oversimplifying their perspective as a "life endures" attitude. In the 10th-anniversary year of 1996, for example, stories in a series titled "10 Chernobyl Years" appeared over the course of several weeks, providing rich ethnographic depictions of how Chernobyl affected life in different contaminated districts within the Gomel region. Yet even in such moments of heightened visibility, individual voices that questioned Chernobyl-related policies were drowned by stories echoing the new government approaches, such as "There is nowhere else we can go from here" or "As long as there are children born in the 'zone,' life goes on."[66] When *GP* raised issues of radiation protection, the discourse of normalization suggested that doses would decrease only when people assumed responsibility for their own health and that the challenge of radiation was one of indifference to its dangers. In this way, individuals—not the state—came to be held responsible for radiation protection.

Somewhat surprising is that these trends were equally true for *EV*, the supplementary publication of *GP* that had devoted much of its early coverage to Chernobyl and had become a national newspaper in 1993, with the State Committee on the Problems of the Consequences of the Catastrophe at the Chernobyl Nuclear Power Plant (hereafter, "the State Committee on Chernobyl") as one of its cofounders.[67] Stories then took an economic and administrative focus, and the absolute number of articles on Chernobyl started declining sharply. Eight out of 10 issues sampled in the year 2000 lacked a single reference to either the Chernobyl accident or its consequences.

Hypervisibility

In contrast to the government-controlled media, the oppositional newspaper *Narodnaya Volya* offered more limited and fragmentary coverage on Chernobyl.[68] Established in 1995, a year after the election of Lukashenko, *NV* soon began publishing articles designed to counteract official discourse on any number of topics, including Chernobyl. Reports in *NV* critiqued the government's handling of Chernobyl, emphasizing radiation safety and health consequences. When the official media began highlighting the economic challenges faced by the affected areas, *NV*'s coverage criticized that focus while also implying that the government economic policies had been flawed. Yet, instead of weighing in on the discussion of the economic challenges of Chernobyl, *NV* coverage called attention to demonstrations known as Chernobyl Path [*Chernobyl'ski Shlyah*], held annually by the opposition on the anniversary of the accident. *NV* contributors drew parallels between the Belarusian government's Chernobyl policies in the second half of the 1990s and the Soviet handling of Chernobyl, when the political regime allowed the authorities to lie and misinform the public. In general, *NV*'s accounts assumed that the regime, government secrecy, and public knowledge were deeply interconnected. *NV* writers thus discussed Chernobyl in the same breath as other political issues, such as the government crackdown on NGOs. In *NV*, Chernobyl and its aftermath was an inherently political question.

NV's dissenting opinions on Chernobyl often came from groups not prominently featured in the government-controlled media. Specifically, civic groups enjoyed better representation in *NV* than in the official media. Numerous articles also featured individual experts, even though these experts seldom spoke on behalf of organizations; they presented their individual opinions and claimed expertise on the basis of their own backgrounds. There were also some consequential absences: as will be discussed later, *NV* included few voices of lay residents of the contaminated areas and offered little description of their experience.

The accounts in *NV* went beyond opposition to embrace resistance as an explicit goal. In the weeks leading up to the annual Chernobyl Path demonstration in 1998, for example, as the official discourse dwelled on normalization, *NV* advertised the protest several weeks in advance of the anniversary date. An article about the Chernobyl cleanup workers losing their status and benefits asked directly, "We want to hear: *Was there an accident or not?* Were there liquidators or not? We want *glasnost* [openness] on this question."[69]

The oppositional discourse attempting to make Chernobyl visible displayed a range of features that could be defined as *hypervisibility*. These characteristics included the heavy use of metaphors and highly symbolic images (such as images of bald children with cancer) and the use of hyperbole and uncorroborated numbers. As we will see below, this discourse extended the meaning of Chernobyl as a symbol. Expressions used in this discourse also suggested that the scope of Chernobyl was greater than had been assumed by the government policies. According to *NV*, for example, the Chernobyl Path was motivated by a concern for the lives of future generations. At the same time, the discourse notably lacked concrete, verifiable empirical data or experience-based descriptions of life in the contaminated areas.

Consider, for instance, the highly symbolic images that appeared around the Chernobyl anniversary in April 1998. An article on "the cost of life" of those affected by Chernobyl pictured a bald, obviously sick child superimposed on an image of the Chernobyl nuclear power plant (see figure 3.3). The image did not illustrate empirical claims or experiences described in the article, nor did it relate to the nuances of the text; rather, it evoked stereotypical connections the reader was likely to make between radiation and cancer.

Images advertising the 2000 Chernobyl Path demonstrations—captioned "We Are the People"—similarly related Chernobyl to a set of emotionally charged images (see figure 3.4). Symbols of radiological tragedy appeared alongside images of political opposition: the white-red-white Belarusian flag unrecognized by the state government, the Constitution, a young person whose face is hidden under a bandanna to prevent identification by the authorities, Japanese paper cranes symbolizing the cancer victims of Hiroshima and Nagasaki, and an icon called *Chernobyl'skaya Bogomater'*. A panoramic picture showed demonstrators carrying a bell "that is a reminder of the horrible tragedy that happened in April 1986."

Figure 3.3

"Payments to People Whom Chernobyl Turned into the Disabled," *NV*, April 29, 1998. The article stated, "Payments to people whom Chernobyl has turned into the disabled are 23 times lower [than they should be], and 'compensation' for the families of perished Chernobylites [cleanup workers] is 25 times lower. Today the price of health lost in Chernobyl is set at 100 dollars, and the price of life of a Chernobylite at 150 dollars."

Emotionally charged, exaggerated language complemented the heavy use of symbols, similarly working to extend the scope of the accident. Chernobyl Path demonstrators argued, and *NV* repeated, "Time does not muffle, but rather makes more acute the pain with which Chernobyl burned every—yes, every!—resident of Belarus." A 2000 resolution signed by a number of scientists, liquidators, oppositional leaders, and Belarusian intelligentsia argued that "the nation is on the verge of extinction."[70] *NV* referred to Chernobyl as "one of the greatest tragedies in the history of humanity," "the Belarusian wound that would not heal," a "black misfortune that is

Figure 3.4
Images titled *"Chernobylski Shlyah* [Chernobyl Path] 2000. We Are the People!"
NV published at least one image from this series weekly in March–April 2000, preceding the April 26 anniversary.

also felt in the uncontaminated regions," and the "world's greatest techno-
genic catastrophe."[71] In 2002, Chernobyl appeared as a symbol of a particular type of catastrophe: "Ecological Chernobyl, economic Chernobyl, and
moral and political Chernobyl are creating a systemic crisis [*sistemny crisis*]
in today's society, threatening us with new misfortunes."[72]

Hypervisibility—with its charged emotional statements, symbolically
loaded images, and other rhetorical tools—appeared as a reaction to the
disappearance of Chernobyl in the official discourse. Sometimes these
accounts used unsubstantiated statistics in an attempt to present Chernobyl as a problem demanding public attention. For instance, a 2002
public statement written in Belarusian and signed by 21 public figures
claimed, "During these years, radioactive death [*radyyacyinaya smerts*] has
taken more than 200,000 people; one quarter of the territory of the country is contaminated with long-living radionuclides—it affects 1.84 million
people, 500,000 children and adolescents."[73] The latter two numbers were
similar to the official statistics used at the time, but the source for the first
number, the number of victims, is unclear. My point here is not that one
number was accurate and another was not (indeed, one of my objectives
in this book is precisely to question how various public statistics presented
by authorities are produced). Rather, it is that the use of such large, specific
figures performed a particular rhetorical function, both similar to and different from using such figures of speech as *millions of victims* and *thousands
dead*. The unsubstantiated numbers can be said to do exactly what "real"
numbers would do in this case: tell their audience how much attention to
pay to the problem. They might also be said to serve as a placeholder for
"real" numbers, produced through more trusted mechanisms, should such
numbers ever become available.[74]

The phenomenon of hypervisibility was not limited to the pages of *NV*.
Chernobyl Heart is an Academy Award–winning documentary released in
2004 that described the humanitarian activities of the Irish Chernobyl
Children's Project International, which organized and sponsored heart surgeries for children in Belarus. According to one of my interviewees, a local
expert involved with the project, there were no data confirming that Chen-
robyl-related radiation exposure resulted in heart conditions at the time
that the documentary was shot. The expression *Chernobyl heart* was thus
used to mark and dramatize the potential consequences of Chernobyl in
the absence of officially recognized data. Indeed, in the absence of official
statistics, in a context in which the official discourse avoids the discussion
of radiological risks, what persuasive tools are available to mark the presence of risks and to mobilize broader audiences and policy makers?

Of course, the official media of the late 1990s also contained occasional emotionally charged, dramatic references to Chernobyl that ran counter to the "life endures" discourse of normalization. Compared to the rhetoric in *NV*'s coverage, however, the use of such language by the official media was restrained, did not share the same political motivation, did not have the same consistency, and was generally drowned out by the domineering discourse of normalization. Yet the very presence of such sensationalism in the official media might be indicative of the challenges faced by those who attempted to call public attention to invisible risks with geographically and temporarily unclear contours. Applying the insights from chapter 1 (where we considered lay articulations of the signs of invisible danger), such sensationalism and excessive dramatization might be signs of trying to find ways to talk about phenomena that have not been precisely articulated. This is, in a way, not unlike waving one's hands when there are no words to describe what is happening.

Chernobyl-related coverage in *NV* certainly was not limited to the production of hypervisibility. Although the coverage was not voluminous, it raised questions about children's health recuperation, radiation safety standards, and the loss of benefits for the cleanup workers. *NV*'s early reports on the government's plans to build a nuclear power plant emphasized the memory of Chernobyl. At the same time, *NV*'s Chernobyl-related coverage did not particularly attempt to include local perspectives or describe the concerns of daily life in the heavily affected areas. Indeed, the overdramatic nature of some of the reporting might have betrayed a particular position— namely, that of the oppositional urban intelligentsia. Residents of the contaminated territories were often described as "hostages"—"of the peaceful atom," "of the state," "of the regime," and "of the tragedy." (The government media referred to these groups as "hostages of the circumstances").

The oppositional media's portrayal of Chernobyl was thus shaped in response to the government discourse. The price of hypervisibility may have been the loss of some sense of the concreteness of post-Chernobyl problems and some sacrifice in credibility. The following comment from an interviewee living in Minsk in 2005 highlights this point: "Chernobyl Path appears as something that is not related to radiation, as an attempt of the political opposition to remind us about themselves.... I don't think they have a viewpoint on Chernobyl's consequences as such. It is more of 'We got beaten so many years ago, the bruises are gone, but we've been beaten, haven't we?'"

Conclusion

Discursive representations of Chernobyl have not been stable but have instead been tied to political struggles and transformations. Three issues stand out in the analysis of Chernobyl's metamorphosis: (1) the uneasy relationship between an economic and a radiological framing of Chernobyl and the extent to which economic framing has contributed to the production of invisibility, (2) the role of scientists in rendering radiological contamination and its effects publicly visible, and (3) the relationship among visibility, hypervisibiltiy, and invisibility.

I have argued that the framing of Chernobyl as first an economic and then an economic-administrative problem has contributed to the invisibility of the accident's lingering radiation—and to the gradual disappearance of the topic of Chernobyl as such. This might appear as a paradox: One consequence of the Chernobyl accident for Belarus was massive radiological contamination, yet the topic of radiological contamination has never been the main focus of the related media coverage. Instead, more than 90 percent of all the articles I found discussed socioeconomic issues.

The economic focus in itself does not preclude visibility for radiological contamination as a public issue. Indeed, serious discussion of radiation protection measures might not be possible without considering broader issues. As illustrated by the debates in the last years of the Soviet Union about the Chernobyl Program, discussions of how to live in the contaminated areas and who should be evacuated were directly tied to questions of funding and the hope for international assistance. But a discussion of multiple economic issues (associated with long-term environmental mitigation, effects of early mitigation efforts, and economic development of the affected areas) should be distinguished from the framing of Chernobyl as a predominantly economic problem that foreclosed public consideration of its radiological and political aspects. The official discourse rendered Chernobyl's radiation invisible by redefining "the catastrophe" in almost exclusively economic terms, overgeneralizing and diffusing images of Chernobyl, and displacing scientists from their role as public experts.

During the last years of the Soviet Union, when radiation appeared more prominently in Chernobyl discourse, scientists and the oppositional intelligentsia played a larger role in defining the discussion. They challenged the "radiosecrecy" approaches of Soviet experts and party leaders and the lack of transparency around the Chernobyl accident and ultimately made more information about the accident publicly available. Belarusian scientists also challenged the Soviet scientists' methodological approaches, revealing the self-serving interests of the Soviet administration camouflaged as objective science. The reporting of these scientific perspectives as political was as important as the data themselves.

When the media (relying on local scientists as sources) could approach radioprotective prescriptions as political rather than simply objective, this helped move these highly consequential but technical issues into the public debate. The public controversy over Soviet standards also assisted in the basic work of educating people and incorporated some technoscientific representations (e.g., names of radionuclides, contamination maps) into the cultural vocabulary. After the adoption of the Chernobyl Program, when the scientific and political controversy subsided, the official media again appeared unprepared or unwilling to treat technoscientific data as anything but objective, emphasizing government bodies as key sources for such data. The reframing of Chernobyl as an economic-administrative issue further solidified the role of government officials as sources, allowing little transparency in or discussions of what data were used and how the data were produced.

Finally, we have considered the production of hypervisibility of Chernobyl (and specifically of the radiological risks) in the oppositional discourse, along with some instances of it in the official media. We might ask how the production of hypervisibility differs from the production of visibility, as in the efforts to make Chernobyl more visible in the last years of the Soviet Union. There might not be a clear line separating the two; the difference may be a matter of degree of using specific, empirical evidence. The production of such evidence requires a significant level of institutional and infrastructural support. As will be discussed in the next two chapters, the oppositional intelligentsia could summon significant institutional support and resources in the last years of the Soviet Union, but not under the conditions of economic turmoil and Lukashenko's regime. Nor did the oppositional discourse in *NV* include the perspectives of the residents of the most affected areas. Paradoxically, this omission of local experiences in an attempt to draw the big picture was antithetical to how Chernobyl was covered in the local newspapers, *GP* and *EV*, which often failed to expand its analysis beyond local experiences.

I have argued that visibility and invisibility (and, for that matter, hypervisibility as a particular kind of visibility) are not absolute but relative categories. The direction of the effort—rendering hazards visible or obscuring them—is relative to other positions. At the same time, I consider incorporating local empirical evidence as highly consequential for the production of visibility of imperceptible hazards and public knowledge about them. (We will return to this question of empirical, intimate knowledge as we consider the institutional side of obscuring radiological risks and effects in chapters 5 and 6.)

Shifts in the official discourse materialized in a multitude of local and national administrative practices, including alterations in radiation safety measurement standards, changes in the status of the contaminated territories, the reshaping of scientific institutions and directions of scientific research, and a general creation of conditions in which radiological problems were unlikely to be brought up. We will see that the discursive transformations described in this chapter were enabled by particular infrastructural arrangements and would have great consequences for the infrastructures of radiation protection and Chernobyl-related research.

4   Twice Invisible

Because radiation is not directly perceptible to the unaided human senses and we do not encounter it as a tangible phenomenon, formal representations of what should be considered dangerous become doubly important in defining the scope of contamination and its risks. By formal representations I refer to standards, categories, and thresholds used in radiation protection. They help us interpret raw numbers by providing a context of what constitutes radiation risks. I also refer to visual maps that systematize quantitative data into graphic representations based on these definitions. These formal representations—including such things as acceptable thresholds of human exposure and acceptable levels of food contamination—are the language of legal and administrative decision making. They also set the general public expectations for what is dangerous.

When the official press first published maps of radiological contamination in the Belarusian republic in 1989, three years after the Chernobyl accident, the sheer fact of their appearance in the official press brought the scope of post-Chernobyl contamination into public discussion. These early maps featured black-and-white hatching patterns that covered most of the country. On later maps, dotted lines traced areas with different levels of radioactivity in particular communities; the numbers inside those lines showed the current readings of background radiation. Though not always easy to interpret, these maps made the scope of contamination visible to the broader audiences of the national press.

By the first decade of this century, contamination maps had long ago stopped appearing in newspapers, but one could find large printed maps of the country's radiological contamination in bookstores (such was the case for the 2001 edition referred to in the introduction; the 2004 edition was the last such printing I ever encountered).[1] Unlike the early newspaper maps, these maps used colors of different intensity to signify zones with different levels of contamination. The maps showed the general patterns of contamination, and they showed 23 percent of the country still contaminated with Cesium-137. Yet by actually going to a community that was marked with a particular color on the map, one quickly discovers the map's inaccuracies in representing local, extremely spotty patterns of contamination. Radiologists from Belrad told me that the map is often useless in working with particular communities for another reason: it says nothing about soil and its transfer coefficient. Areas with comparable levels of contamination on the map might have different soil types. Different soil types have different coefficients of transfer, which predict the contamination levels in plants and, correspondingly, the internal doses of the local residents who consume the agricultural produce. For radiologists who are especially concerned with people's internal radiation exposure, the zoning maps of the country thus exclude one of the more consequential aspects of the radiological situation.

This discussion presents different layers of invisibility and visibility. On the one hand, contamination maps—appearing in newspapers, sold in bookstores, or even posted in local administration offices—make radiological contamination visible to the general public. The mere presence of maps or radioprotective standards adds to the public visibility of the hazard; they bring something concrete to discussions that might otherwise remain abstract. On the other hand, the map is not the territory. Maps, standards, and other formalisms are not perfectly descriptive tools; they necessarily obscure some complexities of the distribution of risk. But in principle they can be more or less descriptive. The very presence of maps matters (as we saw in the previous chapter), but it does not guarantee adequate visibility. This chapter considers how formal representations become less descriptive and can even render already imperceptible risks doubly invisible.

More sensitive and descriptive standards would mean a wider scope of recognized hazards and thus potentially greater mitigation costs, which tends to affect corporate and state interests. Formal definitions of environmental risks are therefore the subject of fierce political struggles or, as Ulrich Beck puts it, "definitional struggles over the scale, degree, and urgency of risks."[2] Concealing the hazard reduces the associated mitigation costs and the number of recognized victims, but it also potentially increases the number of actual victims whose health is affected by the unregulated hazard, even if these victims are unaware of their exposure. One obvious way to reduce the scope of recognized hazards is to increase acceptable levels of contamination or exposure.

This approach is not the only way, as we shall see in the example of successive revisions of radioprotective standards in Belarus, each of which worked to either reveal a broader scope of risks or conceal them. We will see that these revisions and the corresponding changes in approaches to dealing with massive radiological contamination reflected shifting power relations. We will consider how these standards were reconfigured and what work is necessary to make formal representations, and specifically environmental standards, more descriptive and more protective.

How Formal Representations Increase Invisibility

Formal representations, says Susan Leigh Star, are created on the basis of ad hoc, situated descriptions through the work of "abstracting (removing specific properties), quantifying, making hierarchies, classifying and standardizing, and simplifying." This work lets formal representations preserve their meaning even as they travel across time and space; in Bruno Latour's words, they become more mobile and "immutable." At the same time, Star points out, formal representations still have to be used in specific work settings, and thus "instantiated in," or adapted to, each of them. This tension between abstract formal representations and the real-time, concrete work of applying them in practice is what Star calls the central tension of formal representations.[3] It is this tension that produces the potential for erasing radiation risks and other imperceptible hazards.

The politics of formal representations is in how much and what exactly they make visible. In other words, the politics of formal representations is about determining their "degree of granularity"—that is, their specificity and concreteness.[4] It is also about determining which properties of empirical descriptions of reality are emphasized, and which are discarded in the process of abstraction and standardization. Consequently, it is important to understand who has created these numbers, for whom, and why. Whose classifications are used? What is the basis for hierarchical ordering?

The selection of what to include and what to ignore in formal representations has consequences. Categories, standards, and other formal representations carry the influences that have shaped them, and they in turn legitimize some actors and judgments and delegitimize others.[5] These politics become even more consequential when formal representations recede into large infrastructures and become invisible.[6] Particular categories or standards become "naturalized," and the implications of their use have to be articulated. As Star reminds us, the central question for understanding the politics of standards and other formal representations is what has slipped away, "what is left out when formalisms are created."[7]

Taking notice of the tension between abstract formalisms and situated work helps in explaining how protection standards can be used to conceal radiological contamination (and, more generally, how environmental standards can be used to conceal other imperceptible risks). Public recognition of a hazard depends on empirical data describing the hazard. Environmental standards can hinder public recognition of the hazard, effectively serving to conceal it, when they are misaligned with what can be easily measured in practice, under the existing socioeconomic and technoscientific conditions. The scope of recognized risks also shrinks when standards and other formal representations ignore what should be measured in practice to better represent the empirical complexity of the risks—that is, to account for complex and changing patterns of the distribution of the risks. Simply put, standards and other formal representations might be said to effectively conceal the hazard when they are not as descriptive, relevant, and easy to monitor as possible (relative to other existing or suggested formal representations). Standards that look good on paper but are not sensitive to conditions on the ground serve to limit the empirical data used in decision making and impede public recognition of the hazard.

At the core of this chapter, then, is the question of how standards and other formal representations relate to what can be done in practice. This is not a stable relationship; it changes as the environmental situation or the broader conditions change. To better align environmental standards with conditions on the ground might require significant infrastructural resources and much work. It requires the work of "unblackboxing" the formal representations—articulating their inherent or emergent biases and oversight, and this calls for rich empirical data.[8] It also requires the work of creating more usable and less problematic representations that are internally consistent and sensitive to relevant, collectable data. Adjusting the formal indicators to practical conditions of measurement is essential to the process of alignment.

Three Concepts of Radiation Protection

We now turn to the early history of post-Chernobyl radiation protection in the Byelorussian Soviet Socialist Republic and the succession of three concepts of radiation protection that, taken together, provide an example of the efforts to reveal or conceal the scope of post-Chernobyl contamination in Belarus. We will consider (1) the concept proposed by top Soviet scientists in 1988, (2) the alternative concept developed by Belarusian scientists and adopted in 1990, a year before the collapse of the Soviet Union, and (3) the revised concept adopted by the Belarusian government in 1995, after several years of independence. The succession of these concepts illustrates how radical historical transformations allowed for sweeping reconfigurations of the formal representations used in decision making and, correspondingly, for the dramatic fluctuations in the recognized scope of hazards. But let us first consider what preceded the attempts to develop a comprehensive concept of radiation protection.

At the time of the accident in 1986, the Soviet Union had no laws regulating the radiation protection of the general population. Normative acts were limited to the radiation protection of nuclear power plant workers and residents of a 30-kilometer (18.6-mile) perimeter surrounding nuclear plants.[9] The Byelorussian SSR also lacked radiologists and dosimeters of adequate sensitivity. Establishing appropriate radiation protection infrastructures in the republic took several months. Early radiation protection efforts included training several thousand radiation control specialists, devising methods for various types of measurements, and establishing a network of a couple of thousand centers of radiation control at affected collective farms, food processing plants, and other facilities. The state system of radiation control included lines of communication and command; it produced radiological analysis and mapping (but these early maps were not made public).

Establishing elaborate infrastructures was not enough to improve radiation protection in the republic, however, as the participants in these early efforts have documented.[10] Collected data about the levels of contamination were not made public, and they were ignored in policy decision making. Administrators within the system of radiation control made their decisions based on directives from above, which often came from Moscow, rather than on the basis of the available empirical data. Belarusian legal documents often were based on similar documents adopted by the USSR Council of Ministers.[11] Selective approaches to radiological data that ignored many aspects of the complex, overwhelming situation were essential to sustaining the control of the Soviet government and its experts.[12] At the same time, almost every aspect of the process, from collected data to adopted policies that ignored that data, was made invisible by strict secrecy, power hierarchies, and inconspicuous technical standards.[13]

In some ways, the results of this policy making were obviously inhumane. In parts of the Mogilev region—farther to the north from the recognized contamination zone next to the Ukrainian border—people continued living with levels of radioactivity comparable to those around the exploded reactor (evacuations weren't organized until four to seven years later). Residents of the 30-kilometer (18.6-mile) zone around the reactor were resettled to other, still heavily contaminated areas, perhaps in an attempt to limit the scope of publicly recognized contamination and to maintain the region's population levels.[14] High radiation doses received by groups of lay population either were not registered or were concealed. Medical staff was ordered to record understated doses. Laypeople could not officially receive the diagnosis of acute radiation syndrome and instead were given the diagnosis of vegetovascular dystonia.[15] Agricultural production continued even in heavily contaminated areas. Indeed, most of the 1986 agricultural yield was processed for consumption, with the exception of products from the 30-kilometer (18.6-mile) zone. The only place where this produce was not brought to and sold was Moscow.[16]

My point here is not to recount this well-documented negligence but to suggest that these inhumane policies become more difficult to interpret when their description relies on formal representations, as in the following descriptions. Note that to interpret the politics of the standards below, one needs not only some technoscientific knowledge (or at least some explanation and comparisons) but also some awareness of the local context. The issue is not just how the emergent Soviet standards compared to other standards but also what they meant in practice.

On May 12, 1986, soon after the accident, the Soviet government set the maximum radiation dose limit at 500 millisieverts per year for the population in general and 100 millisieverts per year for children under 14 and for pregnant and nursing women.[17] On May 22, 1986, the limit was lowered to 100 millisieverts per year for all members of the public.[18] In 1987, the Soviet Ministry of Health set a new "lifetime" dosage limit, of 500 millisieverts over 70 years.[19] (It is worth noting that all of these post-Chernobyl exposure thresholds dwarf the current international standard of 1 millisievert per year for members of the public.) The same year, Soviet scientists proposed radiation protection measures that promised to remove the need for relocating the populations by providing them with uncontaminated food supplies, especially milk, and administering decontamination measures to enable food production. These measures should have enabled people to remain even in the areas that had particularly high levels of contamination: exceeding 40 curies per square kilometer.[20]

Yet the efforts to supply the affected areas with clean food proved unfeasible. Empty shelves and a food shortage was a widespread problem in the last years of the Soviet Union. The production of uncontaminated food in the contaminated areas also was unachievable, despite complex deactivation measures and restrictions on private farming. The residents of the contaminated areas often had to rely on subsistence farming to supplement their diets and had little or no information about radiation protection measures they could implement. The full volume of agricultural production continued on the territories with up to 80 curies per square kilometer until 1990. In the fall of 1988, Moscow experts proposed a concept of radiation protection that would remove the problem of people living in the contaminated areas without access to uncontaminated food.

1. The Soviet Safe Living Concept

The first concept of radiation protection for the general population was part of the Soviet attempt to develop a comprehensive program of mitigating the consequences of the accident in order to move beyond issuing numerous isolated decrees. The concept was an official document that specified standards for radiation protection, including permissible contamination levels and human radiation exposure, criteria for populations at risk, and priorities for protective measures. The concept was meant to provide "a systematic solution" to the problem of post-Chernobyl contamination, possibly serving as a foundation for further legislative acts.[21] In many ways, the first concept was an extension of the earlier Soviet approaches to radiation protection.

"The Concept of Safe Living in Areas Contaminated after the Chernobyl Accident" was developed by the National Committee on Radiation Protection, headed by Professor Leonid Ilyin and approved in the late fall of 1988. It set a new maximum dose of 350 millisieverts (35 rem) for a lifetime of 70 years. The concept assumed that "the sum of external and internal exposure delivered to a person as a result of the Chernobyl accident" would not exceed this level. Ilyin and his colleagues also argued that doses equal to or less than 350 millisieverts for a lifetime would not have significant health effects—that any health effects caused by the Chernobyl accident would be "in the range of values less than [the] standard deviation of spontaneous levels of the corresponding pathology."[22] Based on these assumptions, further radiation protection measures, including relocation, were deemed no longer necessary in almost all the affected areas of Belarus, Russia, and Ukraine. Implementation of the program—lifting all restrictions introduced in the contaminated areas after the accident—was slated to begin in January 1990.

The concept therefore claimed that people could live safely anywhere, except for the zone of alienation (where the evacuations happened in 1986), and that agricultural production in the contaminated territories did not pose a problem. Additional decontamination measures might be required if the projected doses in particular locations exceeded 350 millisieverts.[23]

2. The Alternative Belarusian Concept

The last years of the Soviet Union were a period of a heightened engagement of the media and civil society. A number of public figures in Belarus, including some leading national scientists, challenged the Soviet handling of the aftermath of the accident. The challengers were emboldened not only by *perestroika* (restructuring) and *glasnost* (openness) but also by the hope for international assistance in dealing with the fallout. They had access to more information than the average citizen and pressured the government to declassify documents related to Chernobyl in 1989, in the process exposing nearly criminal negligence in radiation protection and in the collecting of dosimetric information. To them, it seemed that top Moscow scientists represented the interests of the Soviet establishment; the theoretical assessments made by Ilyin and his colleagues were concealing the truth, the real scope of the consequences of Chernobyl. The Belarusian scientists saw themselves as revealing what had been made invisible by the methodological choices of the leading Soviet experts.

The Belarusian scientists' critique of the Safe Living Concept was concerned precisely with areas where the concept obscured the empirical complexity of the post-Chernobyl circumstances.[24] The Safe Living Concept was a one-step system of making decisions about protective measures that ignored how people accumulated their doses and what could be measured in practice. It lacked effective mechanisms for keeping track of dose burdens. In the words of Vassily Nesterenko, one of the most outspoken Belarusian radiation safety experts, "the Ministry of Health of the USSR left the population without measures of radiation protection, inviting them to take care of their health in unrealistic ways."[25] Ivan Lishtvan, the vice president of the Belarusian Academy of Sciences, referred to it as an "office concept" to emphasize that it could not be realized in practice.[26]

Experts at the Academy of Sciences proposed an alternative concept, which they presented to the government of the Byelorussian SSR.[27] The central Soviet government responded by inviting three World Health Organization (WHO) experts with strong ties to international nuclear organizations to visit the republic and participate in the meeting at the Belarusian Academy of Sciences in June 1989 along with Ilyin and his colleagues. The international experts attempted to convince their Belarusian counterparts to endorse the Safe Living Concept. The WHO delegation proposed an even higher threshold and rejected the possibility of any radiation-induced health effects.[28] As we will see, International Atomic Energy Agency (IAEA) experts continue to refer to the Soviet estimates approvingly.

Despite pressure from the Soviet and international experts, as well as resistance from some local scientists, the alternative concept was approved in 1990, and it later became the foundation for the Chernobyl laws in Belarus (see below).[29] The Belarusian concept was both more cautious in its estimates of radiation health effects and more sensitive to the complexity of the post-Chernobyl situation, including what could actually be done (and measured) in practice. The concept was based on two underlying assumptions: (a) the no-threshold hypothesis of the relationship between dose and effect (i.e., there is no threshold below which radiation doses are safe), and (b) combinatory effects (radiation effects do not always manifest as pathology—radiation exposure can also increase sensitivity to influences of nonradiological factors).

The Belarusian concept recognized the lack of adequate radiological data as well as the impossibility of reliable monitoring of individual doses. Thus, multiple criteria were established to guide radiation protection. These criteria included effective equivalent doses, but also levels of soil contamination, the effectiveness of radiation protection measures, and changes to the structure of morbidity. The concept's authors argued that people should not live where uncontaminated food could not be produced. Limitations on agricultural production and the existing ways of life started in areas with 15 curies per square kilometer of Cesium-137; resettlement had to start there as well. Residents of areas with lower levels of contamination had the right to resettle if protective measures were not effective and annual doses exceeded 1 millisievert (from internal and external exposure, in addition to background radiation).[30]

To summarize, the Belarusian scientists attempted to remedy the misalignment between the one-step Safe Living Concept and the empirical complexity it denied—the multifaceted, compounded distribution of radiation risks. The Soviet Safe Living Concept used only one criterion: effective doses. The practical impossibility of keeping track of such doses was a precondition for the Soviet experts' boldly optimistic, reality-denying predictions that all doses would be lower than a dangerous level. The Belarusian scientists' concept was far more cautious about the nature of possible radiation effects. Their concept included lower thresholds, more criteria, and the kind of criteria that aligned well with what was easily measurable in practice (e.g., levels of soil contamination and the effectiveness of decontamination measures). As a result, this concept revealed a much greater scope of contamination in Belarus and made radiation risks far more publicly visible. Moreover, dividing the affected areas into zones based on levels of soil contamination could easily be translated into contamination maps, which helped visualize the hazard for the public (see table 4.1).

Even though the Belarusian concept offered a more nuanced description that relied on measurements that could actually be recorded, it was not a perfect description of radiation risks. In other words, the formal representations were better, but not perfectly, aligned with the empirical complexity of the contamination. The Belarusian concept faced its own criticisms. For example, soil contamination did not necessarily predict people's doses; there was no consistent correspondence between the two. As mentioned at

Table 4.1
Zoning of Belarus Based on Levels of Radioactive Contamination and Dose Loads on the Population

Zone description	Average annual effective equivalent dose	Density of contamination, curies/km² (kilobecqerels/m²)		
		Cesium-137	Strontium-90	Plutonium-238, -239, -240
Evacuation (exclusion) zone	Territory around the Chernobyl NPP; the population evacuated in 1986			
Zone of primary resettlement		> 40 (> 1,480)	> 3	> 0.1
Zone of subsequent resettlement	Annual dose can exceed 5 mSv*	15–40 (555–1,480)	2–3	0.05–0.1
Zone with the right to resettle	Annual dose can exceed 1 mSv*	5–15 (185–555)	0.5–2	0.02–0.05
Residence zone with periodic radiation control	Annual dose should not exceed 1 mSv*	1–5 (37–185)	0.15–0.5	0.01–0.02

Note: * In addition to background radiation

Source: On the Legal Regulation of the Territories Exposed to Radioactive Contamination as a Result of the Catastrophe at the Chernobyl Nuclear Power Plant (1991 law).

the beginning of this chapter, local radiologists would often dismiss zoning maps while working with individual communities, noting that the coefficient of transfer (which depends on the type of soil and predicts the levels of contamination in food) is what actually matters. Neither the maps nor the concept itself tried to account for the patterns of migration of radionuclides. The Belarusian concept also said nothing about the great variability of doses within communities or irradiation duration.[31] Nevertheless, however imprecise, the zoning maps—sometimes posted in local administration offices or sold in bookstores—made the scope of contamination publicly visible. And they showed that almost a quarter of the country was contaminated with long-lasting radionuclides.

The final version of the concept was approved on December 19, 1990. It later became the foundation for two Chernobyl laws, adopted in 1991: (1) On the Legal Regulation of the Territories Exposed to Radioactive Contamination as a Result of the Catastrophe at the Chernobyl Nuclear Power Plant (hereafter, "the Legal Regulation of the Territories law"), and (2) On the Social Protection of Citizens Affected by the Catastrophe at the Chernobyl Nuclear Power Plant (hereafter, "the Social Protection of Citizens law").[32] These laws translated the zones into levels of compensation and entitlements and laid the groundwork for complex socioeconomic and bureaucratic processes associated with resettlement and the claiming of benefits. The number of people resettled after 1991 is seldom explicitly discussed in the media or in publications related to Chernobyl. It is often mentioned, however, that resettlement and compensation were a massive burden on the state and that the state program of mitigating the consequences (the Chernobyl Program) was never completely fulfilled. The approach soon proved too expensive to maintain.

3. The Revised Belarusian Concept
The political thawing of the last years of the Soviet Union made it possible for Belarusian scientists to reveal a far greater scope of the hazard than had been recognized previously. Yet the hazard's greater visibility meant greater costs. The alternative Belarusian concept was adopted with a hope for international assistance that never materialized. Mitigating the consequences of Chernobyl became prohibitively expensive after the collapse of the Soviet Union, when Belarus faced the economic hardships of transitioning to a market economy. In 1992, 19.9 percent of the overall Belarusian budget was spent on the Chernobyl-related expenses, including resettlement and construction of housing for the resettled. In 1993, the Council of Ministers' decree N57 formed a working group to revise the existing concept of

radiation protection and to create a new plan for the rehabilitation of the contaminated areas. The group's task appeared to be to reduce the recognized scope of contamination and, consequently, to reduce the costs of dealing with the aftermath of Chernobyl.[33]

The working group was headed by Evgeni Petrovich Petryaev, the head of the Department (*Kafedra*) of Radiation Chemistry and Chemical Technologies at Belarusian State University from 1968 to 1994. The group's concept declared an end to the period of emergency measures. As the official media said at the time, one had to learn to live with radiation. Many Chernobyl experts critiqued the newly revised concept; it was rejected by the Parliamentary Commission on Chernobyl and declined at least three times by the Belarusian Academy of Sciences' Council for Radiation Protection. The new concept went through a number of revisions, but its key ideas remained the same, and the Council of Ministers approved a version of it in November 1995. It was not publicized widely, although media coverage at the time clearly signaled changes in the official approaches to Chernobyl.

The revised concept excluded all criteria of radiation protection but one—average annual effective doses, now the only criterion for zoning. It eliminated the two criteria that were easy to use in administrative practices: levels of soil contamination and the effectiveness of earlier decontamination measures. Focusing on people's radiation doses might have been a good idea in principle (one could find high doses in areas with moderate soil contamination), but doses could not be measured easily in practice and had to be estimated by scientists through techniques that themselves were controversial and not transparent to the public.[34] The areas where average doses were estimated to be less than 1 millisievert per year—representing more than 70 percent of all contaminated territories in Belarus—would have no additional protective measures and would lose their social protection. (These average doses corresponded to soil contamination levels of 1–5 curies per square kilometer; see table 4.1.)[35]

Technically, the Petryaev group's concept adhered to the original limit of 1 millisievert per year. But the zone with protective measures and no right to resettle was defined so broadly, with a wide range of 1–5 millisieverts per year, that even average doses there could be close to 5 millisieverts per year. The critics also pointed out that since radiation doses vary greatly, individual members could be receiving much more than 1 millisievert per year, even if that were the average dose in a given location.[36] Higher doses were likely, since significant amounts of basic foods—milk, grains, and vegetables—were contaminated beyond the state's acceptable levels known as RDU-92 (calculated on the assumption of 1 millisievert per year of additional exposure from contaminated food).[37] Since rural residents typically relied on subsistence farming, their food was likely to be excessively contaminated. The critics were particularly concerned about radiation exposure of critical groups such as pregnant women and children, who are especially vulnerable to the effects of radiation.

Although no communities became less contaminated, the scope of contamination was about to shrink. From the critics' perspective, the concept used artificial means, simply redefining the criteria to eliminate protective measures in much of the affected territory.[38] The critics described it as a formalistic approach, since the only criteria left did not allow for keeping track of the actual distribution of radiation risk. As with the Soviet Safe Living Concept, the practical impossibility of conducting the measurements and the nontransparency of expert estimates was essential to what the concept meant in practice: the reduction of the public visibility of Chernobyl-related contamination. In other words, the concept achieved its goal of reducing Chernobyl-related costs by increasing the misalignment between formal representations and the empirical complexity, what could actually be measured in practice. Predictably, it also paid little attention to the health effects of radiation and possible preventive measures.

After the Petryaev group's concept was approved in 1995, it became the foundation for the 1996–2000 Chernobyl Program and justified reductions in Chernobyl-related spending. In 1996, the Council of Ministers had approved the revised list of the areas that required radiological protection.[39] Radiation and social protection measures in the territories with 1–5 curies per square kilometer were being terminated. At the same time, the zoning criteria in the Legal Regulation of the Territories law remained the same, even when the requirements of that law were not being met.[40]

After 1996, the shrinking of the contaminated areas was done incrementally, and village after village lost its status.[41] The drop in 1996 was followed by a less pronounced but steady decline in the number of villages classified as contaminated.[42] Some localities lost their status altogether, whereas others had theirs downgraded. It is not clear to what extent these administrative changes reflected actual radiological changes or were guided by projections and estimates.

The subsequent policies of the state combined laudable radioprotective achievements with a gradual erasing of the problem.[43] The state continued with the planned approach to mitigation, adopting a new version of the Chernobyl Program every five years. The approach in the first decade of this century emphasized "rehabilitation" but with radiological control.

Indeed, radiologists from Belrad and other independent critics acknowledged the state's significant radioprotective efforts.

Another five-year program was adopted for 2011–2015, and yet another for 2016–2020, to some critics' surprise and despite the slow but steady shrinking of the scope of recognized contamination. However, various state figures suggest that the problem has been solved (see chapter 3). Anecdotal evidence suggests that many of the benefits for the affected populations have been cut. The residents of the communities that lost their status as radiologically contaminated stopped receiving their compensation, leaving them to potentially assume that radiological contamination was no longer an issue (see chapter 1). A number of other administrative measures were similarly discontinued in those communities, including gasification (the effort to prevent people from burning radioactive wood in their furnaces), free lunches for schoolchildren (so that children could eat at least one uncontaminated meal a day), and health recuperation programs.

The Invisible Work of Making Visible

The 1995 revised concept provided the government with leeway for shrinking the areas officially recognized as contaminated—or, in the official language, "rehabilitating" these areas. That radiological contamination did not completely disappear in the public eye was, to a large extent, the achievement of a small group of experts from Belrad. The late Vassily Nesterenko, the head of Belrad at the time, recognized that post-Chernobyl contamination was too expensive a problem for Belarus. Yet he nevertheless defined Belrad's goal as a system of total control over individual doses.[44]

Belrad's efforts to create this system illustrate the kind of work required to make radiation more visible, such as achieving a better alignment of the formal representations with one another and with respect to empirical complexity. Two of Belrad's initiatives are particularly illustrative. In 1991–1993, Belrad established 370 local centers of radiation protection (LCRPs), where radiologists—local residents trained by Belrad and equipped with Belrad's meters—tested food samples collected from other local residents. Testing food products allowed Nesterenko to demonstrate unexpectedly high levels of contamination in some "cleaner" communities. Belrad emphasized making the data public and educating local residents about ways to reduce their exposure. At the beginning, the LCRPs were funded by the state and managed by Belrad. With new state approaches to Chernobyl, however, the management of most of the centers was transferred to state organizations (usually the Institute of Radiology in Gomel), and many

LCRPs closed soon after. By 2003, there were only 40 state-managed LCRPs left; Belrad retained and funded 19 centers. Most of the centers were later closed after state inspections in 2007 (more about this below).

Belrad also organized the testing of individuals' internal doses with WBCs (see chapter 1). By 1992, there were 102 WBCs in Belarus, most located in hospitals, but the number of devices decreased in subsequent years. The WBCs were heavy and complex devices whose ongoing maintenance required expertise that the hospitals often did not have. As a result, less than one-third of all state-owned WBCs complied with the state standards of certification. As one physician put it, "The only functioning WBCs are at Nesterenko's."[45] Belrad acquired its devices in 1996, and they were light enough to transport in a minivan, which made it possible to conduct testing in remote communities. The data could then be compared with that of the national *Catalog of Doses*.[46] In 2002, Belrad initiated its Forgotten Villages project after 146 more communities lost their radiological protection status. The internal doses registered by Belrad were juxtaposed with the existing norms and thresholds, and according to Nesterenko, some of the government's decisions were reversed.

These examples show that making hazards more publicly visible requires the grounding of formal estimates in empirical evidence. This in turn means work. Furthermore, much of that work is of the kind that Susan Leigh Star calls "invisible work," as Belrad's efforts to establish empirical measurements and "unblackbox" the existing standards illustrate.[47] It requires building tools and infrastructure and sustaining lines of communication. It also includes the mundane, technical work of maintaining equipment, organizing measuring trips, calculating, negotiating, and fundraising.

Making hazards visible is also the kind of work that must be done by experts recognized by the state and with access to standardized, state-certified equipment and techniques. For example, an essential part of making something visible is realigning the formal representations with one another (i.e., triangulating measurements and standards, assessing internal consistency and the logic of standards). In this case, the work of realignment critically depended on Nesterenko's own credentials, his ability to engage in a dialogue with state regulatory bodies, and Belrad's use of state-certified equipment and techniques. Indeed, Nesterenko's own credentials—as a former head of the Institute of Nuclear Energy and a former chief designer of the mobile nuclear power plant Pamir—have also been instrumental for Belrad's efforts to publicize its data and influence decision making (even if only to a modest degree).

This relationship with state bodies has not been without friction, although in the words of one Belrad leader, Belrad's attitude has been "it's better to be friends than not to be friends."[48] Belrad came under particular pressure in 2001 and 2007, when it was subject to two state inspections. The inspection deemed it unsafe to have contaminated food brought to schools for testing at the LCRPs, even if the children were consuming it on a regular basis at home. As a result, nearly all of the remaining local centers of radiation protection were closed. One radiologist described the experience of the two state inspections as "World War I" and "World War II."[49] Nevertheless, Belrad has maintained its broader operations, still striving to systematize its data (along with relevant data from other sources) and make the information public. Belrad's radiologists share their data with local authorities and carry presentation equipment for educational seminars in their mobile laboratories. Bulletins based on testing reports are published online. And their efforts to make data public have reached beyond the borders of Belarus; almost everyone who writes about post-Chernobyl contamination in Belarus uses this information.[50]

As significant and far-reaching as Belrad's efforts have been, the work of making the hazards known and of mitigating them has been necessarily constrained by infrastructural conditions, including the existing technological capacities and the scope of the required effort.[51] For example, before 2012, Belrad did little testing for contamination by Strontium-90, either in food or as internal contamination in people. Measuring bodily accumulations of Strontium, which is a beta-emitter, is an expensive and complex process. Beta-WBCs are not mobile; the Sakharov Institute of Radioecology in Minsk has one, but it weighs more than 400 kilograms (880 pounds). Food testing for Strontium-90 is a laborious and time-consuming challenge. Consider, for example, the protocol for testing milk, which takes a full day. The process begins with separating skimmed milk from cream (only 3–5 percent of the strontium found in dairy remains in the cream). Belrad must purchase this milk from farmers (who typically do not want to just give their milk away, even for testing), then transport it to Belrad's laboratory in a special refrigerated vehicle. Contamination by Cesium-137 covers a far greater territory than contamination by Strontium-90, and the state infrastructures do test for both.[52] The point, however, is that Belrad's testing is constrained by infrastructural factors. We will return to the shaping of infrastructures in chapter 6.

Conclusion

This chapter began by focusing on the politics of formal representations. It examined the successive revisions of radiation protection standards in opposing directions. The standards were set and revised during a decade of political transformation in Belarus, which accounts for the directional shifts and makes this example unusual and potentially revealing. Exactly how the post-Chernobyl standards were reconfigured provides an insight into the practical politics of environmental standards. It highlights how altering the practicality of standards can change, and potentially limit, the public visibility of a hazard.

My argument in this chapter has been that the central tension of formal representations—the tension between abstract formal representations and their use in concrete practice—has implications for the production of invisibility of radiological contamination in particular and for the politics of environmental standards more generally. This inherent tension can be exploited to conceal or reveal environmental risks. We saw how radiological contamination can be made more visible or, on the contrary, obscured by rearticulating the relationship between formal representations and empirical complexity: what can be measured in practice, under the existing socioeconomic and technoscientific conditions, as well as what should be measured to better describe patterns of risk distribution. These rearticulations can produce either greater alignment or greater misalignment between the formal representations and empirical complexity. We also discussed the infrastructural dimension of alignment: the constant work of monitoring this relationship in order to create more usable, descriptive, and relevant standards. As the complexity and scale of the risks presented by a particular situation increase, so does the importance of the work of alignment.

An analysis of the work of aligning has several broader implications. First, creating more descriptive standards, or adjusting environmental standards so that they remain descriptive and relevant, can be incredibly labor-intensive. It might require not only specialized expertise but also extensive infrastructural work. In the case of the Belrad radiologists, this work included establishing infrastructures, obtaining appropriate equipment and certifications, maintaining equipment, and collecting and analyzing data. This infrastructural work is typically invisible to the broader public, but it is indispensable for challenging existing standards and producing better alignment both among the formal representations themselves and with respect to the empirical complexity of the risk. Some of this work might be carried out only from certain technoscientific and bureaucratic positions; it requires specialized knowledge and tools, along with access to the mechanisms and networks of decision making. But it is never just about expert knowledge. Adjusting formal representations of environmental standards critically depends on local, contextual knowledge.

Second, the work of alignment requires public unblackboxing of standards and questioning of their politics and implications. This analysis has suggested that institutional secrecy provides the conditions for greater misalignment and thus the potential concealment of risks—not necessarily intentionally, but by way of removing structural opportunities for the work of alignment. Secrecy protects the "black boxes" of standards and amplifies their politics. In this sense, scientific controversies help, since they compel the actors to articulate the implications of the existing standards. It appears to be significantly more difficult to discern what happened with radioprotective standards in Belarus in the late 1990s and in the first decade of this century, as the society was becoming increasingly nontransparent. The practical implications of ongoing changes to Chernobyl radioprotective policies are also not easy to ascertain (especially since the names of the Chernobyl laws and the radioprotective norms remained the same). It is not as clear how, and to what extent, the concept of radiation protection is instantiated in practice—that is, exactly how various agencies have been reconciling state radioprotective efforts with state policies of "rehabilitation" and "sustainable development." (Whatever the shortcomings of the radioprotective efforts, those efforts are, as I heard even critics of the state policies say, unparalleled.)[53] This raises questions about the accountability of experts and transparency of decision making, as well as ways to ensure that expert institutions are sensitive to local, contextual knowledge and reflective of the possibilities on the ground, including the possibility of enabling risk monitoring by broader groups of the public.

Third, familiarity with local circumstances, including local socioeconomic and technoscientific conditions, is essential, not ancillary, to setting environmental standards and assessing their politics. This conclusion has much in common with Lawrence Busch's analysis of the requirements for ethically producing standards.[54] Busch argued that the production of standards should be delegated to subsidiary bodies, and my analysis of the post-Chernobyl case illustrates why. Responsible configurations of environmental standards require local, situated knowledge—including the knowledge of what can be easily measured and done in practice. Technoscientific expertise is not enough; the lack of local knowledge, or strategic ignorance of local conditions, is likely to result in formalistic standards and, as a consequence, limited public recognition of environmental risks. Local knowledge is essential to rendering radiological contamination, and other environmental hazards, visible.

The work of articulating and monitoring standards is a consequential yet often publicly invisible part of defining radiological risks. Standards become formalities hiding the scope of imperceptible hazards when they are not descriptive, relevant, and easy to monitor, ideally by a greater number of experts and the public. Functioning standards depend on the presence of adequate conditions for the work of alignment between formal representations and empirical data. Yet even the best standards cannot necessarily make radiological contamination visible by themselves. The next chapter focuses on the rhetoric of some consequential international reports on the health effects of Chernobyl. In both that chapter and the one after, we will return to the juxtaposition of empirical assessment and theoretical estimates as a site for the production of invisibility.

5 No Clear Evidence

Of all the different perspectives on the scope of Chernobyl's health effects, the perspective of the UN nuclear experts has been the most unyielding. According to a September 2005 joint news release issued by three UN agencies that were members of the Chernobyl Forum—the International Atomic Energy Agency (IAEA), the World Health Organization (WHO), and the United Nations Development Programme (UNDP)—"fewer than 50 deaths had been directly attributed to radiation from the disaster, almost all being highly exposed rescue workers."[1] According to these agencies, the accident's only observable health effect on the general population was an increase in thyroid cancers in children. The experts projected a total of 4,000 deaths from radiation exposure, which was, as noted in the press release itself, close to the estimates made by the Soviet experts in 1986. Belarusian, Ukrainian, and Russian scientists often made diverging assessments and found their data on the post-Chernobyl health effects ignored or met with suspicion. In the words of Tamara Belookaya, physician and head of the NGO Belarusian Committee "Children of Chernobyl," "What we [Belarusian scientists] have is a presumption of guilt. We do research, and it is assumed to be wrong until proven right."[2] Even the dramatic increase of radiation-induced thyroid cancer in children, now widely acknowledged, was for several years dismissed by international experts. Critics of the international agencies' approach interpreted this eventual recognition as a sign that there were "no more arguments to reject the reality in this case."[3]

This chapter considers the public stance of the IAEA, WHO, the United Nations Scientific Committee on the Effects of Atomic Radiation (UNSCEAR), and other UN organizations that issued authoritative reports on the consequences of Chernobyl.[4] Their public perspective was not necessarily representative of internal tensions, negotiations, and practices within these organizations. Yet their reports made consequential public arguments, establishing patterns of recognition or opaqueness for the health consequences of a massive nuclear accident. The positions articulated in these reports were reflected in the Western media portrayals of Chernobyl, as illustrated by the following headlines after the Chernobyl Forum press release: "Chernobyl's Harm Was Far Less than Predicted," "Experts Find Reduced Effects of Chernobyl," and "Little to Fear but Fear Itself"; similar headlines appeared as far back as 1991.[5] The reports issued by these influential UN organizations also had implications for Chernobyl-related policies in Belarus, including state radioprotective efforts and state management of research institutes.

The major reports from IAEA, WHO, and UNSCEAR rendered Chernobyl's radiological effects significantly less visible than critics of the reports did. The comparison to the efforts of the critics matters because the production of (in)visibility is relative; the term emphasizes the directionality of efforts, relative to other perspectives, in a particular historical context. As we will see, the Chernobyl reports from these major international agencies strove for an authoritative and definitive scientific view of Chernobyl's health effects. Yet these reports were also greatly constrained in terms of their references and production, in part because they methodologically favored established theories over the complexity of empirical data. One might find parallels to what Michelle Murphy has called "regimes of imperceptibility," in which the design of experiments and methodological approaches makes it impossible to register harmful exposures detected through alternative means.[6]

The international agencies' incredulous view of radiological health effects came in tandem with their view of the affected populations as suffering from anxiety and stress. The poor health of the affected populations was interpreted as a consequence of anxiety and economic hardship and not in any way the result of chronic radiological contamination of their environment and food. Exposure to contingent and uncertain scientific information was often said to exacerbate matters, underscoring the need for authoritative knowledge. The production of invisibility considered in this chapter is thus grounded in a kind of symmetrical simplification of the complexity of radiation health effects and the needs, capacities, and decision-making context of the affected populations. The expert reports on Chernobyl did not attempt to represent the viewpoints of the affected populations, even though the texts often invoke them in justifying their own recommendations.[7] The perspectives of local scientists are also missing, and so is a particular dimension of empirical questions and data that I call *intimate knowledge*: knowledge obtained through context-sensitive, in-depth analysis that acknowledges its own uncertainties. Later international assessments turned to emphasizing the economic consequences of Chernobyl, yet this emphasis only fortified the underlying symmetrical simplification in international reports and the exclusion of the perspectives of local scientists.

The existence of international bodies—including the IAEA, UNSCEAR, and the International Committee on Radiation Protection (ICRP)—relying on the work of expert committees to assess radiation risks was a product of the 1950s.[8] These bodies are fairly interconnected, permitting the circulation of experts among them. The role of scientists within this institutional framework for assessing radiation risks is, in the words of historian Soraya Boudia, "enormous and visible." She also notes that in this institutional context, scientists have often been expected to provide assessment that would help calm down public anxieties and "rebuild public trust." The scientists' expertise provided grounds for claiming objectivity and delegitimizing opposing views. The framework of expert committees removed the debates on radiation risks from the public arena of political debates into a controlled institutional setting, making challenges to established expert opinions "difficult and costly."[9]

Critics argue that the role of WHO has been affected by the 1959 agreement between it and the IAEA.[10] Within the UN system, WHO reports to the Economic and Social Council, whereas the IAEA, though an autonomous organization, reports to the General Assembly and the Security Council. [11] The 1959 agreement established the IAEA's control over what information would be distributed to the public. It stipulated that WHO research programs would be subject to consultation with the IAEA and that their results must not be released if they interfered with the operation of the nuclear agency. As a result, the IAEA, not WHO, took the leading role in estimating the consequences of the Chernobyl accident. The problem with this arrangement, critics have noted, is a potential conflict of interest arising from the IAEA's mandate to promote peaceful uses of nuclear power and to promote standards for nuclear safety.

Chernobyl strengthened the IAEA's role in establishing nuclear safety standards, but it challenged the agency's mission to advance the uses of nuclear power. The accident exacerbated the problem of the public's perception of nuclear power as an unsafe source of energy. Hans Blix, director-general of the IAEA from 1981 to 1997, argued five years after Chernobyl that "the future of nuclear power depends essentially on two factors: how well and how safely it actually performs and how well and how safely it is perceived to perform."[12] The nuclear experts' response to the challenges posed by Chernobyl was to attempt to manage Chernobyl-related public fears. As anthropologist Sharon Stephens described it, the experts sought to affirm their control over the areas of uncertainty and the chaos of reported consequences, to reaffirm "solid scientific grounds for current policies," to present expert opinions as unanimous, and to harmonize disparate national safety standards. Achieving these objectives relied in part on maintaining the boundary between experts and the public, especially the affected population. Rational, scientific conclusions of the experts were juxtaposed with the "irrational, uneducated, emotional, and sometimes even hysterical" public reactions.[13] This approach has had implications for the extent of the international recognition of Chernobyl's health effects.

Reducing to Certainty

One of the more notable IAEA successes in the wake of Chernobyl was establishing an exchange of information with the Soviets following a period of silence after the accident. At the IAEA postaccident review meeting in August 1986 in Vienna, the Soviets delivered a report on the causes of the accident, and the IAEA experts fully accepted the Soviet interpretation of both the causes of the accident and its radiological consequences. The report "was welcomed internationally for the light it threw on the whole incident."[14]

But as Adriana Petryna describes it, the Soviets practiced a selective approach to radiological data: "the Soviet truth (as presented to the IAEA) prevailed above and beyond observable evidence and realities of the plume." Facts that did not support the Soviet response to the accident were either not collected or not acknowledged. Petryna argues that such bracketing of the tremendous and unfolding complexity of the situation was essential for the Soviet government and its experts to sustain their authority and control. A massive accident, a "catastrophe whose scale was unimaginable ... became manageable through a particular dynamic: non-knowledge became crucial to deployment of authoritative knowledge."[15]

The 1986 Soviet report claimed there was no acute radiation sickness among the general population, only among Chernobyl personnel and firemen who responded to the accident. For the general population, the report forecast negligible stochastic (probabilistic) effects—calculating, for example, an increase in cancer mortality as less than 0.05 percent of the existing spontaneous mortality rate from cancer. Later, the local scientists—those from the former Soviet republics and not, for example, part of Leonid Ilyin's group at the Institute of Physics in Moscow—described the 1986 Soviet report as false information.[16] When it was produced, however, the report (and all information related to the accident) was classified in the Soviet Union and it remained secret until May 1989.[17]

As we saw in the previous two chapters, 1989 was a turning point for the public visibility of Chernobyl in Belarus, which was still the Byelorussian Soviet Socialist Republic at the time. Public awareness of the contamination and the extent of the Soviet cover-up exploded. Local scientists disagreed with the proposed Soviet Safe Living Concept for radiation protection, which essentially claimed that additional radiation protection measures for residents of the contaminated areas were not necessary. The Soviet government responded by inviting three WHO experts to visit Belarus and participate in a meeting at the Belarusian Academy of Sciences in June 1989, together with leading Soviet radiation medicine experts. The international experts supported the Soviet concept and even suggested that the threshold for lifetime exposure could be higher.[18]

Their report to the Soviet government rejected the possibility of any observable radiation-induced health effects and included the following statement:

Scientists who are not well versed in radiation effects have attributed various biological and health effects to radiation exposure. These changes cannot be attributed to radiation … and are much more likely to be due to psychological factors and stress. Attributing these effects to radiation only increases the psychological pressure in the population and provokes additional stress-related health problems, it also undermines confidence in the competence of the radiation specialists. This has in turn led to doubts over the proposed values. Urgent consideration should be given to the institution of an education programme to overcome this mistrust by ensuring that the public and scientists in allied fields can properly appreciate the proposals to protect the population.[19]

The statement suggested that the psychological vulnerability of the public was so significant that attributing health effects to radiation might worsen people's health. From this perspective, disagreeing with these expert assessments would in itself create a health risk. This and later Chernobyl assessments by the international radiation experts appeared to treat "psychological pressure" as a concept that required no self-reporting or testimony from the affected populations themselves. The authority of nuclear experts was presumed sufficient for diagnosing populations with anxiety. The methodological challenges of locating the problem "in the head" were not considered explicitly. Instead, locating the problem in people's heads appeared similar to what in science and technology studies is referred to as *blackboxing*; the causes of poor health no longer had to be or could be investigated in the world external to the affected populations but were now fully contained within the individuals themselves.

The summary brochure of the International Chernobyl Project, a major study organized by the IAEA in 1990–1991 at the request of the Soviet government, approvingly reproduced the above quote. The Soviet government's request was part of its ongoing effort to counter the more cautious claims of Belarusian scientists. In October 1989 the Supreme Council of the Byelorussian SSR explicitly rejected the Safe Living Concept, instead adopting a version of the Chernobyl Program grounded in the Belarusian scientists' alternative concept for radiation protection. As Piers Paul Read put it, "it became clear to the Soviet leaders that on the question of Chernobyl they no longer enjoyed the trust of their own people.... The eminence and experience of the nation's leading scientists counted for nothing." The same October, the Soviet government requested assistance from the IAEA to assess the Safe Living Concept and the radiation protection measures undertaken so far. The Soviet state was "virtually broke," and lowering radiation safety thresholds would require more resources for additional radiation protection efforts, including massive relocation from the heavily contaminated areas.[20] With its request for assistance from the IAEA, the Soviet government explicitly sought to boost public trust in the Safe Living Concept.[21]

The International Chernobyl Project involved 50 research missions and about 200 scientists, although it was not meant as a comprehensive, long-term research study. In approaching the Chernobyl data, the project's experts relied on risk estimates derived from the study of the atomic bomb survivors in Hiroshima and Nagasaki that had been carried out by the Radiation Effects Research Foundation (RERF).[22] The RERF data, though based on analysis of a different population and a different radiological context, was valued for providing "considerable knowledge as to what kinds of health effects may occur and what doses may produce them," as well as "information about how long after exposure these effects may appear." According to the project's experts, the Soviet scientists overestimated the doses actually received. Defining radiation effects as cancers and hereditary effects, the project concluded that "on the basis of the doses estimated by the project and using internationally accepted risk estimates, future increases over the natural incidence of all cancers or hereditary effects would be difficult to discern, even with well designed long-term epidemiological studies."[23]

Other health effects were described as unrelated to radiation. The project's experts described local studies' attempts to connect somatic disorders with radiation exposure as "failures" potentially caused by lack of equipment and trained personnel and by poor access to scientific literature.[24] The international experts' "firm grasp," as Petryna puts it, over what constitutes radiation-related health effects, what magnitude of doses can lead to these effects, and how these effects are to be demonstrated provided grounds for their claims to authority.[25] The project ignored incongruous data, such as an account presented to the IAEA Secretariat by the Belarusian Minister of Health that described "a significant increase in the morbidity of thyroid" in heavily contaminated districts of the Gomel region, an increase in the rate of hereditary malformations in newborns, increased rates of some somatic diseases, and a more difficult and prolonged course of many chronic diseases.[26] As Petryna notes, the IAEA experts' readiness to ignore the significance of raw data that did not fully match their methodological requirements and expectations raises ethical questions and questions of accountability.[27] At the same time, the project's experts did note "considerable psychological consequences such as anxiety and uncertainty, which extended beyond the contaminated area." Although the report was careful not to define this phenomenon as radiophobia, it nonetheless argued that anxiety was "compounded by the socioeconomic and political changes occurring in the USSR."[28]

The release of the project's results did not help the UN's first efforts to organize Chernobyl-related assistance. Faced with a dire economic situation, the Byelorussian SSR had issued an official appeal on February 20, 1990, for cooperation and humanitarian assistance in dealing with the consequences of the accident. The United Nations released a joint plan for international cooperation to mitigate the consequences of Chernobyl a year later, in March 1991.[29] The International Chernobyl Project released its conclusions on May 21, 1991; the publication of the full report was delayed until October 1991.[30] But since the United Nations did not hold its Chernobyl Pledging Conference until September 20, 1991, the effectiveness of the campaign was muted by the IAEA's claims of the absence of observable radiological consequences. According to the State Committee on Chernobyl, the Chernobyl Pledging Conference was a "complete failure" and produced less than $1 million, less than 1 percent of the amount requested by the three most affected Soviet republics. The second international pledging meeting, in 1998, collected only $1.5 million—2 percent of the requested amount—$1 million of which was "a targeted U.S. deposit on implementation of Ukrainian projects."[31]

Inside the Public's Heads

Five years after the release of the International Chernobyl Project's report, the IAEA, with the European Commission and WHO, coordinated another major assessment of Chernobyl's consequences: the One Decade after Chernobyl conference, held April 8–12, 1996, in Vienna. The increase in children's thyroid cancer was finally acknowledged as related to radiation, but otherwise the assessment of Chernobyl's health effects remained the same. No radiation-related health effects—interpreted as increased rates of cancer and hereditary effects—were or would be detectable against a spontaneous rate. The reported increase in "non-specific detrimental health effects other than cancer" was considered an artifact of extensive medical examination of exposed populations, though "any such increases, if real, might also reflect effects of stress and anxiety."[32]

A separate section in the summary of the conference results was entitled "Psychological Consequences" and it diagnosed the affected population with "significant psychological health disorders and symptoms ... such as anxiety, depression, and various psychosomatic disorders attributable to mental distress." These psychological effects were said to be caused by economic hardship after the collapse of the Soviet Union as well as by a misperception of radiation risks: "The distress caused by this misperception ... of radiation risks is extremely harmful to people." Expert disagreements about radiation risks were also assumed to adversely affect people's psychological well-being and their psychosomatic symptoms.[33]

The nuclear experts stated the following:

The lack of consensus about the accident's consequences and the politicized way in which they have been dealt with had led to psychological effects among the populations that are extensive, serious and long lasting.... The effects are being prolonged by the protracted debate over radiation risks, countermeasures and general social policy, and also by the occurrence of thyroid cancers attributed to the early exposures.[34]

The summary report of the conference concluded that "the symptoms such as anxiety associated with mental stress may be among the major legacies of the accident." Furthermore, radiation protection measures might only aggravate people's psychological problems and contribute to overall economic decline: "In view of the low risk associated with the present radiation levels in most of the 'contaminated' areas, the benefits of future efforts to reduce doses to the public still further would be outweighed by the negative economic, social and psychological impacts."[35]

Several aspects of these statements deserve commentary. The notion that psychological effects could be provoked by a lack of scientific consensus appears to depend on particular assumptions about the role of scientific assessments in public life as well as on a particular view of the public. It appears to assume, for example, that expert assessments, in and of themselves, create the public meaning of radiation-related issues. This assumption is based, as Brian Wynne has shown in a different context, on interpreting public concerns narrowly as concerns about risks, and more specifically, risks as defined by the experts.[36]

This view of public concerns reduces their scope and ignores their context. At the same time, the nuclear experts' simplified model of the public and the public's concerns also allows these experts to suggest that the affected people's poor health stems primarily from their lack of understanding of the real risks and from the lack of scientific consensus (which in turn legitimizes the calls for institutional mechanisms to ensure "authoritative" knowledge about Chernobyl's consequences). This model of the public molds diverse groups of the affected population into a monolithic group with no economic, educational, cultural, individual, or circumstantial differences that might influence their interpretations of radiation, their anxieties about it, or their radiation-related behavior. Such simplification depends on expert distancing from the affected populations and on the absence of any adequate mechanisms of engaging with the affected populations and accounting for their lived experiences, including those related to health.

This approach has had staying power, although assessments appearing 20 years after the accident also placed greater emphasis on economic causes—in ways that aligned with the earlier IAEA and WHO assessment of Chernobyl's health consequences and built on their model of the public. The UN Chernobyl Forum, organized in 2003 at the IAEA's initiative and engaging a number of other UN bodies, aimed to provide "the most comprehensive" evaluation of Chernobyl's effects to date.[37] According to Abel Gonzalez, IAEA's director of Radiation and Waste Safety:

People living in the affected villages are very distressed because the information they receive—from one expert after another turning up there—is inconsistent. People living there are afraid for their children. The aim of the Forum is not to repeat the thousands of studies already done, but to give them authoritative, transparent statements that show the factual situation in the aftermath of Chernobyl.[38]

This time, according to an IAEA news release, "public information specialists will be involved in the work of the Forum from the outset."[39]

The creation of the Chernobyl Forum followed a watershed report by the UNDP and the United Nations Children's Fund (UNICEF). Their 2002 report *The Human Consequences of the Chernobyl Nuclear Accident: A Strategy for Recovery* emphasized the socioeconomic consequences of Chernobyl and called for sustainable development of the affected regions in Belarus, Ukraine, and the Russian Federation.[40] In a country-specific report on Belarus, the World Bank reiterated a similar perspective.[41]

The new approach emphasized that the accident's environmental and health effects should be considered together with its socioeconomic consequences. Chernobyl disrupted the lives of communities, creating social and economic vulnerabilities: "While physical processes are gradually reducing the level of radioactive contamination in the environment, the most vulnerable groups of people in the affected areas are facing a complex and progressive downward spiral of living conditions induced by the consequences of the accident and the events that followed." The approach was meant to be "holistic," "integrating health, ecological, and economic measures."[42]

Underlying both reports' emphasis on sustainable development was the familiar view of radiological consequences as insignificant and of the public's health as being primarily affected by a misperception of radiation risks.[43] Having been labeled "victims of Chernobyl," the reports claimed, the residents of the affected areas believed that the accident had negative health consequences.[44] This misperception fed into a "victim mentality," an attitude of apathy and fatalism, and a "culture of dependency" on state support.[45] The solution proposed was to build "local capacity" and give "individuals and communities control over their own futures." The public should also be provided with truthful information about the "real risks."[46]

It was in the wake of these reports that the creation of the Chernobyl Forum was announced by the IAEA, which stated that it would contribute to the implementation of the new UN approach to Chernobyl. A major attempt at consensus building, the Forum included representatives of the three most affected countries, along with a number of UN organizations, including the IAEA, WHO, UNSCEAR, and the UNDP.[47] The IAEA working group of experts considered the environmental effects; an expert group convened by WHO reviewed Chernobyl epidemiological studies and prepared a summary report on the health effects.

The experts' assessment of the radiological consequences did not change. Like the authors of previous reports, the WHO group saw no clearly demonstrated increase in the incidence of solid cancers or leukemia in the general population resulting from radiation. No sufficient epidemiological evidence was found for "non-cancer and non-thyroid health effects," although the report of the WHO group did devote a section to them. Statements expressing methodological reservations underplayed the mention of possible increases in leukemia, cataracts, and circulatory system diseases among the Russian cleanup workers; the experts called for more evaluation of competing causes.[48] Generally, the reviewed Chernobyl studies were said to report associations and lack some features (e.g., sufficient control groups or statistical power) that would allow arguments for radiation causation and exclude possible confounding factors. The Hiroshima and Nagasaki data were used for projections and comparisons, whereas the post-Chernobyl epidemiological studies were deemed unlikely to establish causality or to be useful.[49]

In contrast to the methodological skepticism demanding precise and particular demonstrations of radiological causality of cancer or noncancer health effects, no such doubts about the reality of psychological consequences were evinced by the Forum. The experts concluded, without reservation, that there was "a major effect," even though "mostly these mental health consequences in the general population were subclinical and did not reach the level of criteria for a psychological disorder."[50] As evidence, they cited studies diagnosing the affected populations with increased levels of anxiety, depression, and "medically unexplained physical symptoms." According to the Chernobyl Forum, parents were said to pass increased anxiety on to their children "through example and excessively protective care." The affected populations were said to show

a strong sense of lack of control over their own lives. Associated with these perceptions is an exaggerated sense of the dangers to health of exposure to radiation. The affected populations exhibit a widespread belief that exposed people are in some way condemned to a shorter life expectancy. Such fatalism is also linked to a loss of initiative to solve the problems of sustaining an income and to dependency on assistance from the state.[51]

The Chernobyl Forum's booklet asked, "Do people living in the affected regions have an accurate sense of the risks they face?" It answered that "misconceptions and myths about the threat of radiation persist, promoting a paralyzing fatalism among residents." People make unnecessarily negative assessments of their own health because of their fear of radiation and from a sense of "fatalism." Fatalism then causes "both excessively cautious behaviour (constant anxiety about health) and reckless conduct (consumption of mushrooms, berries and game from areas of high contamination)."[52] Residents' difficulties in sustaining an income were also said to stem from this sense of victimhood. According to the Chernobyl Forum, the real problems were poverty and anxiety, not radiation.

The international experts' assessment of the radiological effects of Chernobyl thus remained sanguine. UNSCEAR, which had also produced "authoritative and definitive" reports on Chernobyl in line with the assessments by the IAEA and WHO, concluded, "Lives have been seriously disrupted by the Chernobyl accident, but from the radiological point of view, generally positive prospects for the future health of most individuals should prevail."[53]

The View from the Ground

In contrast to the IAEA, WHO, and UNSCEAR reports, local studies have observed a multitude of health effects related to chronic exposure to low-dose radiation.[54] These health effects include leukemia (not only among the cleanup workers but also in the general population); thyroid, stomach, lung, breast, and other types of cancer; congenital malformations; cardiovascular and blood diseases; neurological diseases; digestive diseases; urogenital diseases; endocrine, immunological, and respiratory diseases; accelerated aging in both adults and children; and increased child and infant mortality. An increase in general morbidity in the affected territories is often noted.[55]

The critics speak of the "catastrophic worsening" of children's health; for instance, the number of "practically healthy" children decreased from 80 percent in 1985 to 20 percent in 2000, with almost no practically healthy children in the most contaminated areas of the Gomel region. Mikhail Malko underscores the great variety of the effects caused by radiation: "ionizing radiation can cause practically all diseases known for humans, including general somatic diseases as well as leukemia and different solid cancers."[56]

Among the vocal critics of the IAEA, WHO, and UNSCEAR reports on Chernobyl are some local scientists who have managed to cross over the language and institutional barriers faced by non-Western–based scholars, publishing in English and thus obtaining a platform on which to refute assessments of the international agencies. Even if they were not as officially endorsed as the reports of the UN agencies, they gave voice to a number of Western and Ukrainian, Russian, and Belarusian critics.[57] Their critiques were not without precedent; they aligned well with the position of Western scholars such as John Gofman, Rosalie Bertell, and Alice Stewart, whose analysis was at odds with the established nuclear agencies' approach to low-dose radiation research.[58] At the same time, established institutional approaches to radiation protection in the West appear to align with the IAEA and UNSCEAR assessments. Alexey Yablokov and his co-authors explicitly note the language divide: their summary review is based on more than 1,000 titles, primarily published in Slavic languages, whereas the Chernobyl Forum report included 350 publications, mainly in English. The Yablokov et al. book and the similarly critical report of the European Committee on Radiation Risks were denounced vehemently by the representatives of the nuclear agencies.[59]

The critics question the approach of IAEA and UNSCEAR to low-dose radiation research—particularly their reliance on risk estimates derived from studies of Hiroshima and Nagasaki bombing survivors. The critics' concern is whether, and to what extent, the risks associated with chronic internal and external exposure to low-dose radiation can be extrapolated from the risks estimated for the Japanese survivors, who were exposed to high acute doses of external radiation. The dominant model assumes a linear relationship between dose and risk at low doses (although some have suggested the existence of a threshold beyond which there is no risk).[60] Critics argue, however, that internal exposure to radiation—that is, when radionuclides are consumed with food and accumulated in the organs of the body—causes greater damage. They also point to the limitations of the model itself. Alice Stewart, a British epidemiologist and an early critic of the Hiroshima and Nagasaki studies, argued that the populations that were studied had already survived the bombing and lived for several years before the studies begun. Stewart's biographer Gayle Greene writes, "This was not a normal or a representative population"; it was a population in which the weakest died off.[61] Because the Hiroshima and Nakasaki studies did not account for what is known as the healthy survivor effect, reliance on that data leads to the underestimation of the possible health effects of irradiation at low doses.[62] Furthermore, the critics point out, comparisons with the old model are used to dismiss new evidence, including evidence from Chernobyl.[63]

International nuclear agencies' research practices have also come under criticism. John Gofman, a nuclear and medical scientist who in the late 1960s had a notable disagreement with the U.S. Atomic Energy Commission over radiation safety guidelines, lists a number of methodological strategies that allow the health effects of low-dose exposure to pass undetected, and he charges the International Chernobyl Project and other studies by the IAEA and WHO with all of these violations.[64] These strategies include, for example, an a priori limiting of the field of study to effects that had already been established. Gofman also lists ways of handling the data that corrupt the validity of scientific results: "the retroactive alteration of databases, the replacement of actual observations by preferred hypotheses, the artificial constraint of equations to rule out certain dose-responses, the subdivision of data until even the largest database becomes inconclusive, and more."[65]

Others have criticized the exclusion of significantly exposed groups, selective dosimetric monitoring, and research designs that disregard local conditions that might shape radiation doses and their effects. For example, Vassily Nesterenko has described an effort by German researchers to measure internal radiation in Belarus that focused mostly on town residents, limiting testing of the rural population—which often accumulates much higher doses—to one village.[66] The International Chernobyl Project was also criticized for failing to detect significant internal radiation doses, as though only uncontaminated food was being consumed in selected villages, and for excluding some affected groups that received high doses, such as early evacuees and 600,000 cleanup workers.[67]

Yet a few vocal Chernobyl scholars who have criticized the IAEA and UNSCEAR handling of the Chernobyl data also have had to respond to the methodological critique and dismissal of the local epidemiological research. Yablokov writes that the IAEA, WHO, and UNSCEAR "demand a simple correlation—a 'level of radiation and effect'—to recognize a link to adverse health effects as a consequence of Chernobyl's radioactive contamination," yet a precise magnitude of individual doses is difficult to ascertain for both historical and radiological reasons (it is, however, possible to estimate collective doses). The international experts also fault the data for having been collected without observing specific "scientific protocols."[68] Yablokov and his coauthors respond, "We believe it is scientifically incorrect to reject data generated by many thousands of scientists, doctors, and other experts who directly observed the suffering of millions affected by the radioactive fallout in Belarus, Ukraine, and Russia as 'mismatching scientific protocols.' It is scientifically valid to find ways to abstract the valuable information from these data."[69]

Finding ways to use the existing data becomes more important with the passage of time, since "much of these data cannot be re-created." Dimitro Grodzinsky and his colleagues point out that rather than conducting comparison studies to test the results, the international experts simply dismiss them. These scholars emphasize that it "is methodologically incorrect to conclude the absence of effects from the absence of data." Finally, the authors point out that the principle of precaution has been ignored by the IAEA and UNSCEAR reports, which make "statements that the risks for human health have been exaggerated" while simultaneously "acknowledging incompleteness of the existing scientific knowledge."[70]

Economic Framing in Theory and Practice

The UN-issued Chernobyl reports in the first decade of the 21st century did not simply insist that there is no clear evidence of radiation-induced health effects. Rather, they framed Chernobyl as a predominantly socio-economic problem, thus vindicating the normalization policies introduced by the government of the newly independent Belarus in the second half of the 1990s (see chapter 3). The new UN policy suggestions similarly sought to "promote a long-term recovery." The Belarusian authorities could now claim that even though some people had been "shocked by the statement of the Belarusian leader that 'all the lands of the country should work for the country,' which several years ago marked the starting point of a new Chernobyl policy," the UN subsequently made recommendations "in unison" with the Belarusian president.[71]

In some respect, the UNDP and UNICEF report and the Chernobyl Forum report pointed toward an even more dramatic course of recovery, arguing that the government could not afford to continue with the current policies, which were already underfunded. Chernobyl-related policies should thus be streamlined and refocused, they said. Instead of maintaining "the sense of victimization and dependency" supposedly encouraged by state policies, the recommended strategies would reverse "the downward spiral of living conditions."[72] Since these reports recognized minimal radiation risks, their socioeconomic recommendations effectively served to further reduce the costly visibility of Chernobyl.

The policy changes recommended by the UNDP and UNICEF report, for example, sought to eliminate most preventative measures. The report held that eligibility for benefits and compensation should be determined on the basis of "actual injury," cases in which the causal connections have been established, rather than "exposure to risk."[73] Chernobyl-related assistance would thus be limited to thyroid cancer patients, as well as to economically vulnerable groups. No mass screenings were deemed necessary; they were to be replaced with targeted screenings. No radioprotective measures were endorsed for territories with contamination levels below 15 curies per square kilometer. Based on the same logic, the UNDP and UNICEF report recommended that programs that tied benefits and compensation to the place of residence—that is, to exposure to increased levels of radiation rather than to proven health consequences—should be discontinued. Health recuperation programs, free meals at schools, Chernobyl-related benefits, compensation, and health-care provisions should all be eliminated or at least reconsidered. Any remaining Chernobyl programs should be tested for cost-effectiveness, including thyroid treatments and the production of "clean" food. In short, the recommended policies would limit Chernobyl visibility in three ways: by narrowing Chernobyl-related categories, by cutting Chernobyl-related programs that had served as the main vehicle for its public visibility, and by generally blending Chernobyl-related problems with broader societal issues (such as mainstream health care or social services). Even international humanitarian NGO projects were prompted to frame their objectives in broader terms not specifically related to health.[74]

By the reports' own admission, the Chernobyl problems thus defined were generic in nature, and countrywide reforms were needed. Although such broad socioeconomic reframing of Chernobyl problems made the specificity of radiological issues less visible, the recommendations did not attempt to address the structural, countrywide problems of poverty and unemployment (especially in rural areas), dependence on the state, and the difficult business environment. It is not clear from these reports how poverty in the Chernobyl-affected areas is qualitatively different from that in the areas without chronic radiological contamination; a more general consideration of the political and economic climate for business development in Belarus is not offered. At the same time, the new economic framing of Chernobyl's consequences was not only generally in line with the normalization policies of the Belarusian government; it also held a promise to revive Chernobyl-related aid on the state level by disassociating such assistance from demonstrable radiological consequences. The World Bank report explicitly suggested that Belarus could approach the donor community with appeals focused on "economic development and improvement in the quality of life of the affected people."[75]

Shortly after the publication of the UNDP and UNICEF report, two international projects adopted this new approach: the International Chernobyl Research and Information Network (ICRIN), which began its operations in Belarus in 2003, and the 2003–2008 CORE program. These two projects inherited the by now familiar radiological assessments and models of the affected population.[76]

ICRIN aimed to satisfy "the information needs" of the affected populations, although informing the public was imagined as a one-way, top-down process: the scientific conclusions of the UN Chernobyl Forum" were to be "authoritatively" compiled and disseminated to the affected population.[77] The information dispatched by the experts was generally assumed to be relevant to the concerns of the affected population. The CORE program sought to "return the sense of control to local people" and to "improve the living conditions of the inhabitants of selected districts by reaching out to the people themselves, helping them to contribute to formulating specific and common project proposals."[78] To this end, CORE solicited suggestions for projects related to health, economic development, radiological quality, and transmission of memory of the Chernobyl disaster. CORE worked in four districts, all with levels of contamination at 15–40 curies per square kilometer.

Though infused with the assumptions of the preceding UN reports, these projects still contributed to the public visibility of Chernobyl by their sheer presence as Chernobyl-related projects. They were also more structurally transparent and relatively more accessible than government bodies, and they engaged in some promotional efforts—including TV appearances by the head of the CORE program and, in ICRIN's case, advertising their Chernobyl.info website on public streetcars in Minsk. These projects, somewhat sheltered by their international status, came close to serving as focal points for the discussion of Chernobyl—though not its health effects—in Belarus in the 2000s. The staff members of the State Committee on Chernobyl were much more at ease answering my questions about ICRIN than they were about the work of the committee itself. Representatives of the Department of Science within the Ministry of Health of Belarus immediately referred me to the CORE program, which, they said, "would be able to tell you more." Both CORE and ICRIN promoted some civil engagement and were more open to new perspectives, including those of the local populations, than were either the international committees of nuclear experts or the Belarusian government officials. The projects' structural openness affected their message.

One of the consequences of this structural openness was some public visibility for perspectives that would contradict or complicate the UN Chernobyl reports. For example, although the ICRIN assessment of the information needs of the affected population discusses the misperception of radiation risks and the "lack of objective information" in the media, it also includes a study conducted by a local NGO appropriating the format of *talaka*, a traditional village meeting.[79] The study makes references to the UNDP and UNICEF approach—"so that we can be quoted," by one researcher's admission—before attempting to weave a coherent narrative from the multitude of perspectives and the complexity of the actual socioeconomic, administrative, and radiological circumstances. For some time, ICRIN's most visible component, its Chernobyl.info website, included not only the Chernobyl Forum's perspective on radiation-related health effects but also some description of the health effects noted by the Belarusian, Ukrainian, and Russian scientists.

In the case of the CORE program, the model of the public that had been radically simplified in the UNDP and UNICEF report had to be adjusted in the context of interacting and engaging with actual populations affected by radiological contamination. CORE sought to reach out to the people themselves, asking them to formulate specific project proposals. Local residents had to learn and follow the rules and bureaucratic language of the program, but the CORE team members, for their part, had to interpret among the perspectives of the local groups, the international representatives, and the state government.[80] Zoya Trofimchik, the former head of the CORE program, acknowledged the friction lines by saying that she understood "the position of the people" but she also understood the international side of the problem. CORE members had to account for the perspectives of the local residents; Trofimchik argued, "I understand these people. I'm from a village myself. My parents would not leave their place either with or without radiation."[81] But perhaps even more important, the CORE team members found themselves being held accountable for their words and actions by the communities they worked with. As a member of the team told me, "We had to look people in the eyes the next time we saw them."

Yet the CORE program worked within an inherited set of constraints. The daily operations did not engage local Chernobyl scholars or reflect these scholars' perspectives.[82] CORE's radiological component emphasized establishing "practical radiological culture." Similar to the Ethos project, which preceded CORE in one of the villages I visited, the CORE program emphasized acquiring "practical skills to reduce the risk of radiation exposure" and stated that residents should learn to "make informed choices resulting in safe behavior." In the words of the CORE director, "People did not learn to brush their teeth overnight, either." Sezin Topçu, who wrote about the Ethos project, observed that this discourse transferred the work of managing exposure to the affected individuals. While the affected populations were encouraged to be participatory and autonomous, they were also left with the task of managing chronic contamination as their individual responsibility.[83] (This work exceeds many families' resources; see chapter 2.) Residents of the affected communities need infrastructural support to successfully mitigate their radiation exposure. CORE did not offer such an integrated, infrastructurally minded approach.

The CORE program lasted for only five years. The Belarusian authorities and the international community hailed it as a success. At first, the radiologists from Belrad were equally optimistic; they reported some reduction in people's doses as measured by WBCs. After the program ended, however, their assessment of its effectiveness was far more subdued; their testing "showed something different, [that it was] not that successful" in terms of lowering internal doses (from the consumption of contaminated produce).[84] From Belrad's perspective, there is little progress in radiological results without the involvement of the local authorities, infrastructural support, and continuous testing and education. After the Core program ended, a former radiologist observed to me, "Everything is quiet. It is as if everything is good."

Intimate Knowledge

Evgeni Konoplya, the former head of the Institute of Radiobiology of the National Academy of Sciences of Belarus, told me that WHO has described the post-Chernobyl circumstances as a unique experiment in assessing the effects of small-dose radiation on a whole population. According to him, however, what is missing from the IAEA—and, more broadly, the Western nuclear experts'—perspective is intimate knowledge of its effects. This intimate knowledge is important because of the complexity and unique nature of the situation itself.

Describing the complexity of the situation, Konoplya referred to different types of exposures (external and internal), different types of radionuclides (alpha-, beta-, and gamma-radiating elements), and the fact that chronic exposure to small doses might have large delayed effects that could manifest over the course of a single generation or in later generations: "When it is an acute exposure, the effects are immediate and then they tail [i.e., abate] depending on the dose. With small doses, there might be no effect at the beginning, but there can be a significant delayed effect."[85] There are also specific biological circumstances to consider, such as the inhomogeneous distribution of radionuclides in the body, which leads to higher doses in different organs or tissues, or in utero exposures, which are particularly dangerous. "These kinds of aspects are not considered," Konoplya argued.

"Nobody knows this," he added. "Nobody knows what the delayed consequences are. We are trying to prove these effects, to let the international community know. Scientists who study Chernobyl's consequences in Belarus and Ukraine have moved much further and understand nuclear accidents much better than even your [Western] scientists do."

The bitterness of the last comment was a reaction to the resistance that this research has met with from the IAEA and from Western nuclear experts:

I talked about it in your [U.S.] energy committee.... I asked them why they were not interested in Chernobyl research. They told me that they were observing the situation from their satellites. I told them that that was rubbish; they were not going to see anything. They could see that there was an accident, that there was an emission, and that some of it went to the atmosphere, but that was pretty much it. You could never see the intimate mechanisms that I showed them on the example of my own scientific data. How could they see that? ... Join in and work with us, there is enough room for everybody to work and learn here.... There are no joint projects with the U.S. scientists. Or, there was one, but the head of the laboratory was Belarusian.

Konoplya offered examples of the lack of understanding, even resistance, by the nuclear experts from state governments and international nuclear organizations. Historian Soraya Boudia has commented on this feature of the debate about the low-dose radiation; she says it frequently shifts back and forth from precise methodological disagreements of a technical order into "a generalized criticism of the nuclear industry and decision-making models within this area."[86] And indeed, John Gofman points out that experts serving on radiation protection committees are typically appointed by the leadership of state nuclear programs or by governments, since programs relating to radiation and its effects are typically controlled by governments. He emphasizes the lack of a system of independent watchdogs.[87] Similarly, several Belarusian critics told me: "He who pays the piper calls the tune [*Kto platit den'gi, tot zakazyvaet muzyku*]. They are afraid for the future of the nuclear industry" and "There is a nuclear lobby there that preaches that nuclear power stations are harmless. There is big money behind it, superpowerful companies."

After reviewing nearly a thousand regional studies on Chernobyl, Tamara Belookaya and her coauthors found that most of the studies do find radiation-induced effects; this review projects a sense of hopelessness in the face of UN statements that "strive to focus attention on methodological issues rather than on solving real problems." Belookaya and her colleagues write, "Unfortunately, scientific authorities are a significant phenomenon, and it is difficult to struggle with them, especially if they are backed up by the powerful and financially secure lobbying structures."[88] In an interview, she was somewhat more hopeful, arguing that the lack of international acknowledgment was temporary. Even more important, she noted that IAEA experts could not affect the reality of the situation and the local research:

As far as I understand, the only consequence for the IAEA is thyroid cancer in children—there are no other health effects after Chernobyl. I take it very calmly now; I have gone through all [attitude] stages before. I'll tell you more. [With thyroid cancer, we were not supported], but time showed everything. We say now that there are reproductive problems and brain tumors, especially in children. There are problems with the cardiovascular system. Time will tell. Gradually, there are more studies coming out, even Russian studies. With cardiovascular system pathologies, they show dose-effect correlations in people living in the affected territories and in liquidators [cleanup workers]. Time will show everything and teach everybody.

In either case, what effect do these IAEA scientists have on us? Do they help us? No. Can they influence us? No. Let them say whatever they want to say. They cannot influence me. They cannot influence the situation itself. They cannot influence what is. Time will tell; it's a good judge.[89]

Alexey Nesterenko similarly maintained that "it is not a problem that the UN doesn't recognize Chernobyl's health effects. Many lost trust in the UN organizations in this respect."[90] According to him, the Chernobyl reports had not affected the scope of the humanitarian aid by international charities, foreign NGOs, and private citizens—including the support these provide for Belrad itself. In other words, the international assistance that was not on a state level was not affected. Indeed, the UNDP and UNICEF report referred to these efforts as "possibly the largest and the most sustained international voluntary welfare program in human history."[91] Yet Nesterenko also observed that the lack of recognition from the IAEA does affect Belarusian state policies.

If the IAEA experts don't have much effect on Belarusian research, it is also not clear what, if any, effect the Chernobyl research from the three most affected countries has on Western approaches to radiation protection. And even though the international organizations might have little direct effect on individual scientists in Belarus, their reports can have implications for the policies of the Belarusian government, including radiation protection measures, Chernobyl-related benefits and compensation, and state support for research institutes and the directions of scientific research in Belarus. The shaping of these research infrastructures is the subject of the next chapter.

Conclusion

The post-Chernobyl increase of thyroid cancer in children ended up being not just observable but also spectacular: the local scientists could demonstrate that this rare condition (with a morbidity rate of 1 case per year in Belarus before Chernobyl) increased dramatically among the most affected populations in geographically recognizable patterns (with 424 registered cases in the first 10 years after the accident).[92] This remains the only increase in morbidity in the general population that is acknowledged as related to radiation by the international nuclear agencies. That this is the only acknowledged health effect could be interpreted as a comment on the methodological requirements put forth by the nuclear committees—they do not appear to be fishing for less obvious effects; more intricate effects seem unlikely to be recognized as related to radiation.

The questions raised by Konoplya, in contrast, focus on more complex, interdependent, and unspectacular phenomena, potentially requiring longer and more involved observation approaches, supported by extensive infrastructures for data accumulation and analysis. The question of radiation causality—whether and what health effects are caused by radiation—has always been, as Gabrielle Hecht tells us, a historical and geographic question, shaped by particular local conditions.[93] Although the consideration of broader conditions that shape the work of international expert committees is outside the scope of this analysis, their methodological requirements do appear in tandem with the exclusion of the more local, situated, intimately involved perspectives.

The IAEA, WHO, and UNSCEAR reports on Chernobyl routinely refer to the local populations' misunderstanding of radiation risks. As Brian Wynne has demonstrated in his work, reliance on such a deficit model of the public understanding of science points to "science's own lack of reflexive openness." He argues that people are astute at "taking up science as means" but are "wary about its ends and interests." The public uptake of science would thus be improved by the development of more diverse institutional forms of science, which would also help make sources of scientific information more "diverse, independent, and context-sensitive."[94] The answer is unlikely to be a unanimous consensus of scientific experts, but rather their greater public accountability and a greater diversity and inclusiveness within the expert organizations, which might ensure that more experts are intimately familiar with the local context. Persistent use of the deficit model of the public understanding of science is problematic not only because it is not representative of what the affected populations know or do not know, but because it is a sign of speaking from a great distance from the affected populations and of actually excluding them from the conversation.

Finally, as the UNDP and UNICEF report makes clear, the human consequences of the Chernobyl disaster are indeed enormous. The opposition between economic and radiological considerations would be a false one. Tetsuji Imanaka, the editor of several collections of articles on Chernobyl that include work from local and Japanese scholars, writes that multiple disrupted lives—the lives of people who had to leave their communities, who lost their jobs, or who despaired about the future—should be recognized as indirect effects of the disaster.[95] Adequate consideration of the social and economic effects of the disaster depends, just as the production of adequate knowledge about radiological consequences does, on sustaining mechanisms of generating such knowledge, which should ideally include fostering the active engagement of local scientists and the affected populations themselves.

6 Setting the Limits of Knowledge

Professor Yuri Bandazhevsky relocated from Grodno to Gomel to become, in 1990, the first rector of the Gomel State Medical Institute, a position he held until 1999.[1] A pathologist by training, he led a group of researchers studying the effects of the internal accumulation of radionuclides—that is, radionuclides consumed with contaminated food products—on pathogenesis in the cardiovascular, nervous, endocrine, reproductive, and other systems.[2]

Bandazhevsky conducted autopsies to show that the concentration of Cesium-137 in vital organs, including the heart, was higher than average in the organism. He found that children residing in contaminated communities had higher dose burdens than adults from the same communities.[3] His research demonstrated that children's organs began displaying pathologies at Cesium-137 levels of 30–50 becquerels per kilogram, a level much lower than what was generally considered dangerous.[4]

Using a different methodological approach, he and his wife, cardiologist Galina Bandazhevskaya, examined cardiograms of children and the concentration of Cesium-137 in their bodies, showing a dose-effect dependence between the accumulations of Cesium-137 and disturbances of the cardiac rhythm in children. Galina Bandazhevskaya explained to me in an interview that the sorts of heart problems observed with low doses may progress and become irreversible with chronic exposure—when children living in the affected areas continuously consume contaminated food products.

In the late 1990s, Bandazhevsky criticized the government approach to mitigating the consequences of Chernobyl and the wasteful spending of the limited state resources for medical research after Chernobyl.[5] This landed him in trouble. As one local scientist put it, "The only impact his message had was on prosecutors" who began hounding him. Bandazhevsky was arrested in 1999, and in 2001 he was convicted of taking bribes and sentenced to eight years in prison. Amnesty International declared him a

Prisoner of Conscience and demanded his release, relating his arrest to his critique of the Belarusian government's Chernobyl policies. Bandazhevsky was conditionally released in 2005, and later he left the country, his ability to work in Belarus appearing impossible.[6] Two years after his release, when an interviewer asked him who the leading Chernobyl researchers in Belarus were, he answered, "I would like to know that myself."[7]

Although Bandazhevsky's treatment was extreme, it does illustrate the government's commitment to the policies of rehabilitation of the areas affected by Chernobyl. The case of Bandazhevsky, whose work and imprisonment made him arguably the best known Chernobyl scholar in Belarus, also sets the context for analyzing the historical emergence and then disappearance of opportunities for research on the biological effects of radiation in Belarus. My focus will be not on the explicit harassment of researchers but on disruptions to the institutional and infrastructural foundations of research.

The history of radiological research in Belarus is largely coextensive with the history of Chernobyl-related research, which emerged here in the last years of the Soviet Union and the first years of independence. A number of research institutes were established in 1987–1992, mostly in the capital, Minsk: the Institute of Radiobiology, the Institute of Radiation Medicine (with branches in Gomel, Mogilev, and Vitebsk), the Institute of Radio-ecological Problems, the Institute of Agricultural Radiology (in Gomel), and the Sakharov International Institute of Radioecology, which provided radiological training. Several other research institutes had departments or laboratories that also conducted radiological research.[8]

This systematic development of radiological research capacity was part of the state program to mitigate the consequences of the accident. The new state policies of normalization and rehabilitation (introduced in the second half of the 1990s) brought changes to the organization of radio-logical research in Belarus. Certain research directions were encouraged and others discouraged; leadership positions changed hands; and capabilities for data collection and analysis were maintained, added, or dropped. An even more dramatic reorganization of research institutes and priorities took place in the first decade of this century. For studies on health consequences of the accident (one of several key areas of Chernobyl-related research), changes in state policies resulted in a loss of qualified personnel and some subject populations and caused disruptions in data accumulation (a process exacerbated by existing problems with Chernobyl-related databases and classification categories). This in turn increased the likelihood that empirical data would be unavailable and that researchers would focus on theoretical assessments that are not as reflective of the complex realities of the Chernobyl effects.

The absence of adequate infrastructural conditions for data collection and analysis constrains what observations about radiation effects can be articulated in the course of conducting research, thus creating particular areas of public invisibility of Chernobyl's consequences and ultimately areas of ignorance. The restructuring and refocusing of Chernobyl-related science in the 1990s and after 2000 disrupted the infrastructure of scientific data collection and analysis, with consequences that are potentially irreversible. The most pronounced result has been the disappearance of the category of radiation health effects as an object of knowledge, and a dearth of scientists who would publicly claim expertise in Chernobyl-related research.

By suggesting that infrastructural transformations were consequential for the development of Chernobyl-related research in Belarus, I do not mean to question the integrity of the scientists conducting this research. Instead, I am observing how the government has shaped research in the context of pervasive and chronic environmental contamination. The work and achievements of individual scientists should not be underestimated.[9]

Science in the Flux and the Chronic Disaster

A major effort to establish local capacities for research on the consequences of Chernobyl began in the Byelorussian Soviet Socialist Republic (BSSR) in the last years of the Soviet Union. Soviet science, as sociologist Gennadi Nesvetailov describes, had "a degree of scientific self-sufficiency and some isolation from world science, although [it] … was on the periphery in intellectual terms to Western centres of world science."[10] Within the Soviet science infrastructure, Moscow had the largest concentration of research institutions and programs, the most technical equipment and resources, the most qualified scientists, and the leading academic journals.

The geographic distribution of Soviet nuclear research followed this pattern, with Ukraine's nuclear research program the second most developed.[11] Belarusian research on the consequences of Chernobyl thus began from a position on the periphery. The Belarusian republic had no nuclear power plants of its own, and its participation in the Soviet nuclear research program was marginal. (One project to construct a mobile nuclear power plant was headed by Vassily Nesterenko at the Institute of Nuclear Energy.) State secrecy, especially the fact that all research related to radiation was classified, added to the exclusiveness of radiation-related research. Tamara Belookaya noted, "Belarusians were not in the loop."[12] Studying the consequences of Chernobyl's fallout and mitigating its effects would require building a research infrastructure from the ground up.

But despite the uneven organization of Soviet science, the Belarusian republic still had a considerable capacity for science and technology through its own Academy of Sciences, as the result of a limited political commitment in the Soviet Union to foster the scientific base of the constituent republics.[13] After the Chernobyl accident, these research capabilities served as a foundation for establishing new institutes and departments specializing in radiation medicine, radiobiology, and agricultural radiology. The Academy of Sciences drew on the personnel and resources of the existing institutes. Some of the figures that emerged in these efforts came from related fields.

Consider, for example, the role of Evgeni Konoplya (1939–2010), who at the time of the accident was one of the few researchers in the Belarusian Academy of Sciences with expertise close to the topic of radioactive iodine. His earlier research had been on methods of hormone therapy, chemotherapy, and radiation therapy for breast cancer; from 1965 to 1980 he worked at the Institute of Oncology and Medical Radiology of the Ministry of Health. Konoplya proposed the creation of the Institute of Radiobiology to Nikolai A. Borisevich, the president of the Academy of Sciences of the BSSR; when the Belarusian and Soviet Academies of Sciences and Councils of Ministers approved the decision in 1987, the Institute of Radiobiology was established with Konoplya as its director.[14]

Evgeni Demidchik (1925–2010), whose research established the radiation-induced nature of the rise in thyroid cancer in Belarus, was particularly well-positioned for this post-Chernobyl work. He had been conducting research on thyroid pathologies since 1972, and he served as the head of the Department of Oncology at Minsk State Medical Institute from 1974 to 1996.[15] Children with thyroid cancer, rare in the republic, were typically sent to Demidchik. In 1990 he initiated the establishment of the Center for Cancers of the Thyroid Gland under the Ministry of Health.

The last years of the Soviet Union also witnessed efforts at independent radiological research and radiation protection. In 1990 Vassily Nesterenko (1934–2008) established the Institute for Radiation Safety "Belrad" and became its first director. Nesterenko had previously served as the director of the Institute of Nuclear Energy of the Academy of Sciences of BSSR (1977–1987) and the chief engineer of the mobile nuclear plant Pamir (1971–1987). He was part of the Belarusian committee of experts formed in early May 1986 to assess the situation, and under him the Institute of Nuclear Energy conducted an analysis of soil samples and created early maps of the scope of contamination. Nesterenko was later released from his duties as the director because of the "alarmist" letters he wrote to the BSSR government in which he attempted to warn the government of the scope of radiation fallout in the republic and the need for radiation protection measures. In 1990, he became the chairman of the joint expert committee on nuclear energy and radiation protection for Belarus, Ukraine, and the Russian Federation, a position that he held until 1994.

The Program for Overcoming the Consequences of the Catastrophe at the Chernobyl Nuclear Power Plant (known as the Chernobyl Program) was adopted by the republic in 1989, and it included provisions for developing scientific research.[16] A separate section of the program described "systematic planned research" for the purposes of developing measures of "minimization of the consequences of the accident."[17]

The Chernobyl Program outlined scientific research and development in four key areas: (1) studies of radioactive contamination of the environment (including genetic, physiological, and biochemical consequences of this contamination); (2) technologies and methods of agricultural production under conditions of radiological contamination; (3) the effects of radiation on human health, as well as methods of diagnostics and treatment; and (4) technologies for mitigating radioactive contamination, including methods and technologies of radiometric and dosimetric control.[18] Scientific research was envisioned as systematic, planned, and coordinated.

By the mid-1990s, Belarus—now a newly independent country—had sufficient personnel working in the field and had achieved scientific results in all key research areas. By 1996, eighteen institutes of the Academy of Sciences and more than twenty scientific and educational establishments were participating in the implementation of the Chernobyl Program.[19] By that time, research on post-Chernobyl health effects in the affected populations included epidemiological studies, studies of the effects of radiation on the body's functional systems (endocrine, immune, cardiovascular, and reproductive), and research on the combined effects of radiation and nonradiation factors on health of the exposed populations.[20]

The development of such an extensive system of radiological research was accomplished despite the extraordinary challenges of the early post-Soviet period. After the collapse of the Soviet Union at the end of 1991, Belarus's already difficult economic condition worsened dramatically. Spending on science decreased significantly (before generally being restored in the second half of the 1990s); lack of funding made it difficult for research institutes to maintain their equipment and their professional activities. The number of employees in science and technology fields decreased by half from 1990 to 1994.[21] Official descriptions from the period note that the majority of research teams working on Chernobyl-related issues were not paid salaries but continued to perform their tasks "on a voluntary basis."[22] Nevertheless, during this period, which featured severe economic crisis and political turbulence as well as greater political openness, much was achieved—from establishing scientific institutions and programs of study to obtaining some key results (including the demonstration of radiation-induced thyroid cancer in children).

The government assumed more active control over the development of science and technology after the 1994 election of President Alexandr Lukashenko, the adoption of a new Constitution, and the reform of executive power.[23] The new state leadership still deemed government control and planning necessary for the long-term progress of science. The government repeatedly attempted to optimize its management of scientific research and of technoscientific development—adjusting, for example, the hierarchy and responsibilities of science-related administrative bodies.[24] All areas of Chernobyl research were coordinated by the State Committee on Chernobyl, following the outline in the five-year Chernobyl Program. Annual research proposals were approved by the Academy of Sciences and the State Committee on Chernobyl.[25]

As a relatively small and economically struggling country with limited resources for scientific research, Belarus faced particular challenges in managing and optimizing its scientific capabilities. Nesvetailov uses the concept of peripheral science to describe the general priorities and tensions of scientific development in Belarus in the first years after independence. In economically developed and politically dominant centers, "national research" overlaps with "world research." In the periphery, a significant commitment to research can be sustained only in a few selected fields, and national research in those fields is affected by research in the center.[26]

The key issue for a peripheral country like Belarus, then, was the direction of technoscientific research amid limited resources. For Nesvetailov, writing in 1995, it appeared "inevitable" that the science and technology policies in Belarus would refocus on more applied research, prioritizing short-term objectives.[27] Indeed, Belarus, as a newly independent country, faced questions about the general orientation of Chernobyl research, the role of the state in shaping this research, and its relationship to international radiological expertise. Furthermore, in the case of Chernobyl's consequences, science appeared to be closely related to the hope of international assistance and cooperation. The state appeared to have little incentive to fund scientific research just for the sake of science, especially when this science was not recognized internationally, when such research might discourage rather than attract international assistance, and when it sustained the economically costly national visibility of the Chernobyl problem.[28]

What Nesvetailov correctly predicted as a growing emphasis on applied research that would justify its costs was, paradoxically, at odds with the nature of the consequences of the Chernobyl accident. In Ravi Rajan's terms, the consequences of Chernobyl were emergent problems of a "chronic disaster."[29] They were not revealed in their entirety at any given moment; rather their scope emerged, and continues to emerge, gradually. Studying the consequences of chronic disasters requires stable, continuously sustained research infrastructures.

However, the emerging, evolving nature of Chernobyl's consequences also made it difficult to estimate the necessary scope of scientific effort and to sustain it over time. The prospective assessment of the required resources may have been particularly difficult in the absence of a tradition of radiological research predating Chernobyl; as we have seen, radiological research in Belarus had few accumulated institutional resources and entrenched practices.

As a result, research on the radiological effects of Chernobyl proved particularly vulnerable to the state government's attempts to optimize the organization of scientific institutions and put science in the service of its policies of rehabilitation of the Chernobyl-affected areas.

Restructuring Chernobyl Research

Less than a decade after the establishment of the Institute of Radiobiology, Konoplya expressed his concern that support for research on the radiological effects of Chernobyl was on the wane. In 1996 he argued, "Unfortunately, the situation in the Republic today is such that we are not able not only to broaden and deepen the studies, but even maintain them at the same level."[30] He pointed to vanishing opportunities for targeted screening of the affected populations (performed by trips of doctors traveling to the affected areas) and even the dissolution of research collectives.

The new state policies of normalization and rehabilitation of the affected areas were echoed in a number of both gradual and dramatic changes to the structure and agenda of research institutes, which in turn transformed what data were collected and from which populations, what databases were maintained and how, and what kind of research was ultimately possible.

The transformations that were under way at the Institute of Radiation Medicine are instructive. This institute, established in 1987 under the

Ministry of Health, served as a lead organization for studying the health consequences of Chernobyl.[31] The institute studied mechanisms of radiation-induced damage, conducted dose monitoring, forecast changes in the health effects of the population, and provided policy recommendations.[32]

The institute had an outpatient clinic in Minsk and an inpatient clinic in Aksakovshchina near Minsk (established in 1989), which provided health care specifically for the Chernobyl cleanup workers, the evacuees, and the population of the affected areas. In the mid- to late 1990s, the institute's structure and function experienced multiple revisions (leading one former employee of the institute to refer to it as the "Institute of Political Medicine").[33]

Changes to the institute's leadership were one sign of political struggles. In 1993, Vladimir Matukhin, the founding head of the institute, was replaced by Aleksandr Stozharov. Just three years later, Stozharov too was removed from his position. According to one critic of the Ministry of Health and its approaches to post-Chernobyl research, Stozharov's removal may have been related to his opposition to the reorganization of radiation medicine in Belarus and to the revised concept of radiation protection (see chapter 4). On December 17, 1996, the Institute of Radiation Medicine was transformed into the Institute of Radiation Medicine and Endocrinology—even though, as the same critic cited above pointed out, the country already had a developed network of endocrinology centers, departments, and laboratories.[34] Another researcher, who interpreted these changes as related to the "official" position that the only health effect definitely linked to Chernobyl radiation was thyroid cancer, told me, "All radiation medicine has become about endocrinology. I wonder if it is going to be just endocrinology soon."[35] Nor was the policy of deemphasizing radiological research limited to a single institute. In 1999, the Sakharov International Institute of Radioecology was renamed the Sakharov Ecological University.[36]

During this period of reorganizing, the Institute of Radiation Medicine and Endocrinology also lost some of its laboratories. Personnel and equipment alike were transferred to other institutes. The Mogilev branch of the Institute of Radiation Medicine became the Institute of Ecology and Occupational Pathology. The outpatient center in downtown Minsk, originally located close to the railway station (making it easier for patients from outside Minsk to reach it), moved to a different building, farther away from downtown and the railway station. According to a former physician from the center, "They [the officials] told us the building was in unsafe condition.... I think it was all done to fault Chernobyl science. There was no renovation in the original building after we were moved from there."[37]

More transformations followed after 2000. A presidential decree issued on April 14, 2003, relocated all Chernobyl-related institutions to Gomel on the premise that such research should be concentrated "in the most affected area." This approach was in keeping with the new policy of rehabilitation of the affected areas. Researchers also faced new requirements of "practical value" and "economic and social efficiency." The government emphasized scientific support for agricultural production, even in the newly reclaimed areas; it was hoped that scientific findings could offer guidance on growing techniques that would minimize contamination and thereby keep produce in line with the existing norms for radiation protection.[38]

The new policies streamlined directions for Chernobyl research into three subject blocks, each with a corresponding institute located in Gomel. The Institute of Radiology—the former Institute of Agricultural Radiology—conducted research on the rehabilitation of the contaminated territories and agricultural production techniques. The Center for Radiation Medicine and Human Ecology, Gomel, became the head institution for medical research on the consequences of Chernobyl (the Institute for Radiation Medicine and Endocrinology in Minsk effectively ceased to exist in the spring of 2003). Long-term radiobiological and radioecological consequences were the responsibilities of the Institute of Radiobiology, relocated from Minsk.[39]

Developing a research base in Gomel, closer to the most affected areas, could certainly have aided studies on the consequences of Chernobyl. At the same time, the decision to relocate all research to Gomel involved tremendous infrastructural costs, including the loss of qualified personnel who remained in Minsk and the loss of continuity in data collection and analysis. Moreover, Belarusian research institutes generally were concentrated in Minsk (the capital), and many physicians had left the Gomel region after Chernobyl.[40] The move also brought changes concerning which groups of the affected populations were observed and what methodological approaches were used. In at least some areas of Chernobyl-related research, this loss of continuity in terms of data collection and analysis resulted in the shift away from empirically based research and toward more theoretical calculations.

According to Evgeni Konoplya, who relocated with the Institute of Radiobiology to Gomel, the institute had 165 faculty members while in Minsk. Two years after the relocation, it had only 75. Mostly two groups of researchers were willing to relocate from Minsk to Gomel: young scientists and faculty members nearing retirement age who, as a former physician from the Center for Radiation Medicine told me, "might not have been able to find work elsewhere."

Other experienced personnel stayed in Minsk, thereby leaving Cher-nobyl-related research and patient observation. According to one scien-tist who stayed, the government's decision to relocate Chernobyl science "might have been adopted without realizing what it would mean. There are big names in Gomel: Konoplya, [Eleonora] Kapitonova [the former head of the Center for Radiation Medicine and Human Ecology in Gomel] ... but losing all that faculty is a loss of knowledge." Konoplya explained to me:

We have to hire new faculty, they have to learn the research methods, it's a com-plicated and long process. Everything was set here, we had well-trained faculty. Ev-erything has been smoothed out during these years, from 1987 until 2003, about fifteen years, not [the] full seventeen years, since it took time to organize everything in 1987, as well. Much has to be recovered now. We have new, young faculty, they have to get trained until they enter science [voidut v nauku].

But two years after the relocation, Konoplya remained optimistic: "We already have first dissertations and first conference presentations, so some things are happening faster than I had expected. And overall, what we have is a serious research complex."[41]

Research on radiation effects appeared to be discouraged at institutions outside Gomel. One faculty member at the Sakharov Ecological University told me in 2004 that she and her colleagues had to remove any references to "radiation effects" from their research project proposals: "One can study radiation only in Gomel now, where there is a special center for that. Here, one can study chemical or physical processes, as long as they are not radio-logical." (Even though she no longer did research on explicitly radiation-related topics, she still feared that the university itself might be relocated to Gomel.) Another scientist commented on similar expectations for research proposals at the National Academy of Sciences (as the Academy of Sciences was now called).

The Center for Radiation Medicine and Human Ecology, now the lead organization on studying Chernobyl's health effects, opened in Gomel in 2003 after more than a decade of construction of its building.[42] The cen-ter—with both an outpatient center and an inpatient clinic—was com-monly described as a state-of-the-art facility. I was told repeatedly that it had exceptional equipment and capabilities, complete with an attractive hotel for visitors from outside Gomel that even the residents of rural areas could afford.

The expectations for the center were in line with the state reorientation of science toward practical results and economic efficiency. In 2007, when the minister of health introduced Aleksandr Rozhko as the Center's newly appointed head (replacing the former director, Eleonora Kapitonova, and then the former acting director, Elena Sosnovskaya), the minister noted that the center was also supposed to know how to earn money: "You have good ophthalmology—so create regional ophthalmology and earn money."[43]

Whatever the virtues of the building's state-of-the-art features, transferring all medical research on the implications of Chernobyl to Gomel and closing the leading institute in Minsk had negative implications for sustaining the continuity of data collection and analysis and for maintaining the necessary level of faculty expertise. Vassily Nesterenko offers an example. Until the research reconfiguration, the Institute of Radiation Medicine and Endocrinology in Minsk had produced the *Catalog of Doses* of the population of Belarus, which listed the current annual exposure doses for the residents of the affected communities and informed the decision making about the necessary radiation protection measures required for these communities.

When the institute was closed, the team of experts assessing dose burdens also stopped their work. In 2004, when the time came for another edition of the catalog, its production was delegated to the Center for Radiation Medicine and Human Ecology. Nesterenko writes that "the accumulated experience was lost, and new experts proposed to create the catalog based on the proportional relationship of annual exposure doses from the density of soil contamination with cesium-137."[44] This approach theoretically derived annual doses and made little use of empirical data describing, for example, actual internal doses in various communities. The center thus generally estimated lower doses of population exposure; by its calculations, only 220 villages required radiation protection measures. The National Committee for Radiation Protection of Belarus created a group of experts who, using Belrad's WBC testing results for 250,000 people, demonstrated that the annual doses estimated by the Gomel center were 40 percent lower than the actual measurements, Nesterenko reports. The Ministry of Health accepted most of the data, and the new catalog included 775 localities as requiring radiation protection.

The Belarusian State Registry of individuals affected by the Chernobyl accident is also now maintained by the center in Gomel, assisted by seven regional departments of the registry.[45] Collection of data for the registry began in 1993; according to the official records, by 2005 it included 1.7 million individuals.[46] The data is collected through annual medical examinations, which provide the foundation for health-care provisions for the affected groups.[47]

In the early 1990s, some of this screening was conducted by teams of physicians traveling from the Institute for Radiation Medicine in Minsk, but that practice was soon discontinued because of a lack of funding.[48] According to a former physician of the institute in Minsk, the institute's database included data on every child observed by Minsk physicians since the establishment of the institute. It is not clear to what extent this (and other) medical observation data was transferred to and used in Gomel; according to the same physician, "Nobody has asked for it, nobody needs it in Gomel.... All the statistics used to come here; what kind of information comes there is difficult to tell. There is no information now."

Considering the many problems with the current health care system, the idea that most of the affected population receive annual medical examinations—especially when those examinations are done in outpatient centers (*poliklinika*) and not by designated teams of physicians—is a "bureaucratic fiction." A physician from the former outpatient center of the Institute of Radiation Medicine and Endocrinology, who used this term in her interview with me, recounted what the annual examination might entail:

I called a local *poliklinika* , inviting children to come and be examined here, since it is still free and we don't have much of a workload. They tell me, "What are you talking about? We submitted our annual data reports for clinical examinations [*dispanserizaciya*] a few months ago." I ask them, "Have you examined everybody already?" They laughed [and said], "We examined some people and wrote in some more people." The staff there [in the local outpatient centers] is overloaded dealing with acute problems.[49]

In the continuing absence of dedicated screening programs for the affected populations and of reliable general health care, other physicians and scientists I spoke with generally agreed that in the words of the same doctor, "they would see few patients at the Center for Radiation Medicine in Gomel; the rest were going to have to cope at home however they could."

The infrastructural challenges brought on by the reorganization of Chernobyl-related medical research exacerbated the existing problems with the completeness of the Chernobyl registry and the politics of Chernobyl-related categories used for data collection. For example, the Chernobyl registry does not include all the cleanup workers, one of the most exposed groups.[50] According to Tamara Belookaya, 300,000 children were born in 1986–1987, but only 2,500 are in the registry.[51] Another researcher pointed out to me that there is also no separate medical category "children of Chernobyl" in Belarus. Children are particularly vulnerable to radiation exposure; presumably, those who were exposed to the highest postaccident doses prenatally or as children are at more risk of developing radiation-induced effects than those exposed as adults.[52] Without a "children of Chernobyl" category, the group of adults who were children at the time of the accident are statistically invisible, and since they are no longer under age 14 and thus "children" according to the general classification, they blend into the larger pool of potentially less affected adults. The term "children of Chernobyl" itself is commonly used by the media and humanitarian organizations to refer to those who are children *now* and are exposed to significantly lower doses than the earlier generation was.

Defining categories too narrowly can also render radiation-related health effects invisible. Establishing and maintaining consistent categories for reporting health effects—categories that are also uniform across various departments and institutions—requires significant work.[53] Although some efforts have been made to ensure quality control of the information sent to the Chernobyl registry, the exact scope and nature of these efforts is not clear.[54] Tamara Belookaya argued to me, "There are no good statistics; everybody is doing them their own way. I was looking through reports in a hospital in the Gomel region. I looked under 'hereditary development defects' (*poroki razvitiya*), and it listed six cases. I looked under "hereditary anomalies of development" (*anomalii razvitiya*), which is the same thing, and it listed thirty cases." She went even further in characterizing the problems surrounding the epidemiological analyses of the health effects of Chernobyl:

Statistics are corruptible. They can be turned whichever way, and it completely depends on who is doing it. Say there is a girl, [and] she was twelve at the time of the accident. She grows up, and it turns out she has endocrine problems. Later she has a stillbirth and then two miscarriages. Then finally she gives birth to a baby girl who has health problems. But who would ever count her or any of her problems if, for example, she lives in Minsk? Where would she be counted? *There are no statistics, only anecdotal cases.* That's the sense of the games with statistics. When we talk to people here, they tell us, "We are dying off." This one lost her brother, that one has a disabled husband, but nobody can do anything.[55]

Belookaya demonstrated an awareness that political and socioeconomic conditions affect what becomes visible and what can be counted. She gave me several examples from her Russian colleagues, related to, for instance, the job and social security conditions in a Russian town with military uranium production where jobs had multiple benefits and "becoming sick" meant losing these benefits and having no security.[56] Another example illustrated how the visibility of genetic effects was erased when a town did not have a local orphanage. Children with genetic problems were more likely to be put in orphanages, so in a town without an orphanage, the children had to be distributed to other places, which helped erase the numbers.

  Thus, research and data collection on the consequences of an imperceptible, chronic, low-dose hazard with delayed effects requires significant infrastructural resources that are sustained over time. It requires funding streams to sustain them, as well as dedicated work efforts (see "The Invisible Work of Making Visible" section in chapter 4). Furthermore, the direction of these efforts may also become misaligned—or even in conflict—with the government's (or other responsible bodies') changing interests and policies over time.[57]

  Compared to international assessments, the Belarusian state recognized more Chernobyl-related health effects—such as in providing "disabled of Chernobyl" status for those who could claim that their health has been affected by the accident. At the same time, the Belarusian Committee on Connection with Radiation Exposure, which oversees individual claims on the relation of radiation to the individuals' health effects, also follows notoriously stringent definitions of which diseases have been induced by radiation. Even two Chernobyl experts, Vassily Nesterenko and Ivan Nikitchenko, had to struggle to get their own health problems recognized as being related to their extensive past exposures.[58] Excessively restrictive, discontinuous, and inconsistent categories are just some examples of the infrastructural conditions that potentially obscure health effects that could be made observable otherwise.

Reframing Chernobyl Research

New state approaches to Chernobyl became reflected not only in the organization of Chernobyl-related institutions but also in their agenda. Galina Bandazhevskaya commented to me that Belarusian science "is not connecting" health effects—that is, morbidity in highly contaminated areas—to radiation (as Yuri Bandazhevsky had done in his research).[59] Bandazhevskaya is not the only researcher who told me that "nothing concrete" is published on Chernobyl health effects in the medical journals and that the "effects of the radiation factor are described in passing, without conviction." Radiation is associated only with thyroid cancers and no other risks, and research questions are asked about unspecific ecological factors that might lead to health problems in the contaminated areas of the Gomel, Mogilev, and Brest regions.

Evgeni Konoplya has observed the same tendency to explain health problems in the most affected populations with references to ecological factors. According to him, radiation is considered to a lesser degree, and references to broader environmental factors do not clarify how these factors, including radiation, interact. Konoplya argues that "such synergetic action [of different factors together] is too serious. The effects are not just additive, but instead aggravated in this case."[60] Clearly, radiation is not the only thing affecting the health of the populations, even in the most affected areas, and it would be beneficial for Chernobyl-related research to take wider issues into consideration. Yet the kind of research that Konoplya criticizes, with its emphasis on tendencies rather than causes, does little to clarify the exact mechanisms of influence. The ecological approach blends all possible negative influences on health indiscriminately, producing not a better understanding of the complex situated reality of factors and effects but instead an invisibility of the radiation factor.

Bandazhevskaya summarized the situation as follows to me: "The money is spent, the programs are completed, but conclusions just prove what they have to prove. The practitioners, physicians, say, 'We see unusual things, but studying them is the prerogative of scientists.' And scientists say, 'We've looked, but it is hard to single out the radiation factor here'."

Bandazhevskaya connects this practice of not singling out the radiation factor to a fear of being snubbed by the IAEA. "The IAEA is a powerful organization against weak Belarusian scientists," she asserts. "Belarusian scientists won't be able to do much without the help of international researchers." As a solution, she proposes launching international programs for basic scientific research on the effects of low-dose (especially internal) exposure to radiation. According to her, this research would collect evidence and provide the grounds for organizing preventative measures for people living in the contaminated areas (as well as, presumably, inform approaches to treatment). The current lack of scientific results affects what preventative measures are adopted, and, Bandazhevskaya notes, "The person's life is not endless." In her opinion, radiation could and should be treated as a risk factor in public health issues, in the same way as, for example, smoking or obesity. She argues, "When a person is treated, radiation exposure should not be forgotten. Patients, for example, can be checked for their internal accumulation of radiocesium with whole body radiation counters."[61]

A brief look at the national reports on the consequences of Chernobyl and the mitigation of those consequences illustrates how this sidestepping of causal explanations plays out. The reports have been prepared by a collective of a few dozen scientists and administrators and published by the State Committee on Chernobyl. These analytical reports provide an overview of the existing national research and administrative data on the topic. Reports from different years consistently emphasize that the consequences of Chernobyl extend well beyond radiological effects—they also include significant social and economic consequences. Yet in their discussion of Chernobyl's health effects, these reports primarily describe trends.[62] The discussion of causal explanations is far more cautious and appears to be in dialogue with assessments by the IAEA, WHO, and UNSCEAR.

These reports describe increases in all classes of diseases, including somatic diseases, some type of cancers, and congenital defects.[63] The radiation factor is mentioned most explicitly in connection to children's thyroid cancer, but in 1996 it also appears in the report's section on hereditary defects (written by Gennady Lasijuk, whose studies on the topic were sometimes referred to by the Belarusian scientists I interviewed).[64] The 2006 report also refers to radiation in connection with the significant increase in breast cancer among relatively younger women of the Gomel region.[65]

The reports' position on international research on the effects of Chernobyl is somewhat ambivalent. Reports from the 1990s and the first few years after 2000 criticize international experts for minimizing or ignoring the consequences of the disaster. The 1996 report argues (without mentioning a specific organization) that "there is a tendency for underestimation of the catastrophe consequences, bringing it down to the ordinary [nuclear power plant] incident."[66] The 2001 report addresses WHO's reluctance to acknowledge radiation-related health effects after Chernobyl: "With the example of the thyroid gland cancer [in children] it became clear that the risks of morbidity and mortality in case of radiation impact are undervalued, especially at chronic exposure to small doses."[67] Similarly, the conclusion to the 2001 report refers to UNSCEAR's assessment as a "vivid example of a prejudiced attitude to the consequences of the Chernobyl catastrophe." The statement is unequivocal: "Based on arbitrary selection of data, on individual publications, which included practically no works of Belarusian scientists, the [UNSCEAR] report treats incompletely and pretentiously the post-Chernobyl situation in the three affected states."[68] The 2003 report argues that "the perception of the existing problems of the Chernobyl disaster" by "the international community is not fully adequate to their real size and importance."[69]

Despite these explicit criticisms, the later national reports increasingly echo the language of international assessments in their discussion of causality. For example, in the 2001 report, two groups of factors are said to affect health after Chernobyl: radiation factors (including both external and internal irradiation) and nonradiation factors, which are described as "social, economic factors; stress; [and] risk perception" (no other ecological factors besides radiation are mentioned). The mechanisms of the influence of the nonradiological factors are not explained, and the report points out that "It's not quite clear today ... how long it will take to prove the radiation origin of ... pathologies [other than thyroid cancer in children], and to conduct an impartial assessment of the risk factors which may not be directly connected to the radiation effect."[70] The 2003 report expresses great expectations for the UN Chernobyl Forum to produce "proof" of Chernobyl's effects and to overcome the disagreements in estimating Chernobyl's consequences; these disagreements, it says, continue to cause "difficulties in attracting international assistance." In similar language to the UN reports on Chernobyl described in chapter 5, this national report also mentions the "Chernobyl victim" syndrome, which supposedly prevents the "active involvement of the population in social and economic activities"; it similarly recommends education programs to mediate the misperception of radiation risks. The misperception of radiation's danger "leads to persistent psychological discomfort."[71]

Similar comments—that the misperception of risks and "psychological discomfort," along with radiation, affect people's health—appear in the 2006 report. It generally emphasizes the rehabilitation of the affected areas, both as a focus of state policies and as a new focus of international cooperation and assistance. The report mentions "a combined influence of radiation and non-radiation factors of the Chernobyl accident" but never considers the question of causality in more detail.[72] After all, a better estimation of the radiological health effects, as well as of the synergetic effects of radiation and other factors, might be counterproductive to obtaining international cooperation and assistance (see chapter 5).

Whereas the national reports produced by the State Committee on Chernobyl appear to be highly conscious of the international assessments of Chernobyl's consequences, individual scientists appear equally concerned about the directions of state policies. After Yuri Bandazhevsky's imprisonment, his Gomel school of researchers collapsed; according to Galina Bandazhevskaya, "There are studies, but no enthusiasm." Scientists are explicitly aware that the government is no longer investing in Chernobyl research. One scientist from the Sakharov Institute told me, "We've known for a long time which way the wind is blowing." Another researcher, who continues working and publishing on Chernobyl-related health effects, called my attention to a series of media interviews with Yakov Kenigsberg

(the head of the National Committee on Radiation Protection and a representative of the government of Belarus in UNSCEAR) in which Kenigsberg
flatly denied that any other diseases, besides thyroid cancer, were caused by
the Chernobyl accident.[73] A physician from the former Institute of Radiation Medicine and Endocrinology in Minsk also told me:

I have heard [the head of the State Committee on Chernobyl] on TV saying that we
are a poor country, so the programs are going to fold.... I would never have believed
that we would stop working [on Chernobyl] so soon. We thought that we would
have enough Chernobyl problems to last us for a hundred years.[74]

In April 2009, journalists writing for the *Vecherni Grodno* newspaper con-
veyed a similar impression of the directions of the post-Chernobyl research
in Belarus:

In the studies related to health effects, Chernobyl has become an unpopular topic....
We heard that for several years now dissertations on this topic are considered un-
promising. We reached the director of the Center for Radiation Medicine and Human Ecology Aleksandr Rozhko, who used to work in Grodno. Yet he refused to
sound any topics of scientific research. Another direction of research is riding the
wave at the moment—how to return contaminated areas into use.... Eleven such
pilot projects started for the period 2006–2010.[75]

In 2012, my experience was similar to that of *Vecherni Grondo*'s journalists:
Aleksandr Rozhko refused my attempts to interview him about research at
the center.

The health effects of the Chernobyl accident have become increasingly
obscured in scientific research in Belarus, both as a result of the conscious
politics of the state management of science and, as Matthew Cresnon put
it in his study on the "un-politics of air pollution," as a result of scientists "taking cues."[76] The perception of the position of those in power can
be a determining influence even in the absence of actual intimidation. It
appears that Belarusian scientists also "take cues" from administrative discourse; explicit directives to reframe research might not be necessary.

Missing Experts and the Radiation Factor

Despite multiple infrastructural challenges, including the reorganization
and refocusing of science related to Chernobyl radiation, there is little
doubt that in some theoretical and applied areas, Belarusian scientists have
accumulated unique experience, what Konoplya referred to as "intimate
knowledge" of Chernobyl's consequences (see chapter 4). It is not surprising that the clearest examples come from thyroid cancer research and

treatment, where accumulated theoretical knowledge (which is also sensi-tive to the local context and the particularities of local cases) translated into better standards of care.

A physician who was one of Demidchik's colleagues highlights the unique Belarusian experience and knowledge by contrasting cases of chil-dren who have been treated in Belarus with those whose parents took them for treatment in the United States, Germany, or Sweden, motivated by their belief in the power of Western medicine and medical technologies. The cases show that "Western doctors simply don't know these patients" in terms of the projected courses of their illness, risks, and optimal treatment. This also holds true for specific issues of thyroid cancer and pregnancy: "these types of problems are a rarity there [in the West]. For us, it's rou-tine."[77] Outcomes for children who underwent treatment abroad compare poorly to patients treated in Belarus, she explains:

One case is the mother who arranged for the child to be operated on and treated in Germany. They even removed his thymus and did not give him the right treatment. When they were in Aksakovshchina, the mother asked me why her son is doing so badly compared to other children with the same original problems. I asked her, "Why did you take him to have his surgery in Germany?"

Examples from this physician also illustrate that the lack of observation of health problems doesn't mean the absence of health problems. Screen-ing for thyroid problems was cut because of lack of funding; as a result, thyroid problems are usually discovered early only in rare groups still rou-tinely examined, such as pregnant women and (ironically) young medical students:

Those who were small children at the time of the accident are particularly at risk for thyroid cancer; they are now young people ages twenty to twenty-five. They come to us as students and find out their own thyroid problems. And many young women, since all pregnant women get tested, and their thyroids are checked. Some of them are pretty far along. We currently have thirty such patients. We have made about twenty-five to twenty-six surgeries after sixteen weeks of pregnancy.... If you don't do a surgery, the cancer can spread.

Adequate screening might not even be the biggest problem with the current state of Chernobyl-related science in Belarus. Despite dedicated research institutions and centers (and clear leadership in understanding and treating pediatric thyroid cancer), there are now very few publicly and professionally visible, publishing experts who publicly claim expertise on the health consequences of Chernobyl.[78] Some of the most prominent experts who resisted the growing invisibility of Chernobyl's consequences

died in the first decade of this century: Vassily Nesterenko, Evgeni Konoplya, Ivan Nikitchenko, and Tamara Belookaya, among others. Some, like
Yuri Bandazhevsky, left the country. A few others, while continuing their
research, prefer to keep a low profile. Galina Bandazhevskaya works as a
pediatrician and is no longer involved with radiation-related research. During my first interview with him, Nesterenko asked me if I could help an
established Chernobyl scholar who had lost his job. A number of scientists
and administrators I attempted to speak to in Minsk claimed to be "not
competent" or "no longer" involved with this field. Top officials at the Ministry of Health advised me to talk to the director of the international CORE
program, claiming lack of relevant expertise.

Radiation as a morbidity or mortality factor has also been rendered less
visible. Depending on personal views, most scientists I approached on the
topic of Chernobyl either claim that "nothing has been found" (and, for
example, that "radiation effects on the cardiovascular system seem to be
too far-fetched") or describe anecdotal evidence to prove the reality of
Chernobyl effects. For example, on two separate occasions, different experts
told me the story of a known female surgeon who worked in the Gomel
region and later was diagnosed with potentially radiation-induced cancer.
The invisibility of both experts and radiation effects is further exacerbated
by the relative information vacuum and lack of space devoted to scientific
discussion of the effects of Chernobyl. Only one annual publication was
devoted to Chernobyl, and several researchers commented on the lack of
Chernobyl studies in regular medical journals.[79]

One scholar still working on Chernobyl told me, "It is all so clear here—
all the consequences are very clear—but it is too bad that the science is
so political." Another researcher insisted, "There is good, fundamental science, but everybody is very cautious about it. It is received very carefully."
Belookaya, when asked whether she felt hopelessness because of the difficulties in making the effects of the accident visible or recognized, replied,
"Often. Pretty badly."

Conclusion

What we know about the consequences of an invisible, pervasive, and
chronic environmental hazard with delayed health effects cannot be
assumed to be a straightforward reflection of the extent of contamination or
the severity of its effects. This knowledge depends on infrastructures of data
collection and analysis, on observational practices, and on already established knowledge. Obviously, then, this raises the question of what kinds of

health effects, under what kinds of circumstances, can become knowable. What conditions (i.e., features of disease and its incidence, structural conditions of research, methodological approaches, and, perhaps, political savvy of researchers) can make the link between radiation and particular effects become visible?

Belarusian research on Chernobyl's consequences followed an up-and-down trajectory. In its current state—after the loss of faculty, when the observed populations have shrunk, and when previously collected empirical data about doses and effects are no longer available—research runs the danger of becoming less empirically based and empirically sensitive, as we saw in the example of estimating dose burdens. As Konoplya said at a Chernobyl conference in 2005, "Theory and practice are inseparable, but practice introduces its own corrections"; the turn away from empirical data obscures the complexity of radiation health effects. Furthermore, scientific assessments of the consequences of Chernobyl determine the scope and extent of radioprotective efforts, treatment sensibilities, and specialized treatment programs—and this means that both our knowledge and the ongoing reality of the consequences of the accident are in a kind of interaction with infrastructural conditions for research, radiation protection, and opportunities for articulating this knowledge.

The Belarusian government's efforts at rehabilitating the affected areas are reaching into the heart of the contaminated land: in 2011, President Lukashenko announced the revival of agricultural production in the areas immediately adjacent to the "zone of alienation"—that is, the Polesski State Radioecological Reserve. Indeed, "rehabilitation" has not been an entirely empty promise. Infrastructural solutions implemented by the state helped lower the doses experienced by resident populations (as measured by Belrad radiologists). These efforts included state radiological control of food processing, free school lunches, health recuperation programs for children, and, in many areas, gasification (see chapter 2). More recently, providing clean pastures for cow gazing has eased the problem of contaminated milk, one of the main sources of high doses in some areas. In 2010, the government extended the Chernobyl Program for five more years; many people had believed that it would not be extended because the Chernobyl problem in Belarus has officially been fixed.

New reasons have emerged for rendering Chernobyl increasingly invisible. The Belarusian government has been considering the possibility of building its own nuclear power plant since the 1990s. The plans were revived again in 2006 and became definite in 2011, after the government obtained a credit from Russia and signed a construction contract with a foreign wing of the Russian nuclear corporation Rosatom. Despite strong objections from Lithuanian authorities, the protests of environmental groups, and public disapproval at home, the nuclear power plant will be built in the northern region of the country, near Ostrovets. In the words of one ecologist in Minsk, "The country most affected after the Chernobyl accident is building its own nuclear power plant. The IAEA has just gotten a great argument," referring to the supposed insignificance of chronic exposure to low-dose radiation.[80] One might assume that the Belarusian government is not interested in making the establishment of post-Chernobyl health effects a priority in the national and international arenas.

Conclusion

Clear, visible clues about possible actions and their potential consequences help us navigate the use of various objects we encounter; we tend to run into trouble when our environment does not make connections between possible actions and outcomes obvious.[1] Radiation is a treacherous hazard because it gives us no clue to its increased levels in the environment, in our food, or even in our bodies; nor do we get any sense of what the consequences of staying in such an environment or consuming such food might be. One of my interviewees, a researcher at the Sakharov Institute of Ecology, emphasized the problem of delayed consequences with an example that "nobody is jumping in front of a car—it is clear how that would end." In contrast, the extent of the danger from radiation must be made visible to us. My argument in this book has been that imperceptible hazards such as radiation can either be rendered more publicly visible and observable or be increasingly obscured, depending on how they are represented. Consider the contrast with the societal treatment of microbes, another imperceptible threat. So much infrastructure has been built around protection from microbes that they are tremendously visible to the public as a hazard; its presence is attested to, for example, by antibacterial soap or signs in restaurant bathrooms reminding staff members to wash their hands. A different picture emerges as we trace the representations of post-Chernobyl radiation risks in Belarus and as we consider what infrastructural conditions exist for articulating these risks or sustaining their visibility.

Much of the analysis in this book has focused on how radiation risks in Belarus were ultimately rendered less visible to the public. The nearly complete public disappearance of an invisible hazard, radiation risks, is not unique to the post-Soviet context of Belarus. Countless other imperceptible hazards, including those in the West, are continually being made invisible by the industries that produce them, and that in turn are aided by administrative bodies that do not regulate them. The tobacco industry has infamously worked to make the health effects of smoking publicly invisible. The chemical industry has been waging a campaign against recognizing the health and environmental effects of pesticides—and it waged a personal campaign against Rachel Carson, who did so much to make the risks of pesticides visible in her seminal book, *Silent Spring*. Historical and sociological studies have documented the various strategies used by industries to displace dangerous toxins as objects of public attention: reframing the public debate on hazards, promoting fake debate where there is a scientific consensus, silencing critics, orchestrating studies to counter even strong evidence of harm, blaming victims' genetic makeup or lifestyles and denying environmental influences, and presenting a lack of monitoring as an absence of health effects.[2] These strategies appear in cases of hazards created as a result of accidents as well as in cases of routine production of hazards. Indeed, even climate change is a complex phenomenon that cannot be perceived directly, that needs to be made publicly visible, and that some interests are trying to make invisible.[3]

In all these cases, the production of invisibility is relative; the term implies the comparison of different perspectives, some of which render the hazard more publicly visible and observable. The dynamic of the public recognition of an imperceptible hazard is subject to power relations; we cannot assume that public knowledge about hazards and protection standards just constantly improve. Public visibility depends on whose voices can be heard and which groups have what kinds of institutional and infrastructural support. What is distinctive about the disappearance of the consequences of Chernobyl in Belarus is not their disappearance per se but rather the trajectory of that disappearance—the historical waves of invisibility and visibility of Chernobyl's consequences. What stands out especially is the eruption of the visibility of radiological contamination in the last years of the Soviet Union, with the corresponding media coverage, passage of laws, and establishment of research institutes. Official recognition of the vast scope of the hazard was made possible by the collapse of the old political regime, with its power relations and its expertise, and by the hope of international assistance.

The influence of the Western nuclear industry is everywhere in the Belarusian case: in the lack of adequate international assistance on the state level, in the international studies that worked hard to find nothing, and in the reports that ignored the research of local scientists and blamed the affected populations' lifestyles or their fear of radiation. Although the organizational structures and practices of the international committees of nuclear experts that produced reports on Chernobyl are outside the scope of this study, it appears that the work of these committees has been constrained by their institutional agenda and a lack of independent experts. In contrast to their critics, these expert committees appear to simply not have been looking for Chernobyl health effects beyond strictly delimited methodological and theoretical frameworks.

Over time, the public invisibility of a hazard can become irreversible in the absence of adequate infrastructural support for data collection and analysis and for the articulation of risks, or when such infrastructures are disrupted. Invisibility then turns to ignorance. As we saw in chapter 6, the restructuring of key research institutes, the disruption of studies and databases, and institutional pressure on researchers to reframe their topics undercut earlier efforts to establish research capacity and practically erased attention to radiation as a risk factor in the health of the affected populations. The consequences of Chernobyl have become an area of nonknowledge most directly through the disappearance of local scientists who would claim expertise in radiological research. In the absence of adequate conditions for data collection and analysis, empirical research on the consequences of Chernobyl is then replaced with theoretical assessments, which does little to challenge the dominant conceptions about the lack of radiation effects. Radiation effects dissolve into individual health problems of nonspecific origins. Thus, before we can assess the number of victims of Chernobyl put forward by the international nuclear committees or their critics, we should consider what social conditions ensure the sustained production and consideration of the local empirical data. The problem is not just with a potential underestimation of the affected populations but also with compromised or nonexistent conditions for data collection and analysis.

Like scientific assessments of the potential harm of radiation, radiation protection efforts require adequate infrastructural conditions. Making the equipment for testing widely available and producing empirical data is the first step. Yet enacting protection measures also requires infrastructural solutions. Imperceptible hazards such as radiological contamination are chronic and pervasive in the environment. For affected groups that lack adequate infrastructural support, reducing radiation exposure from a contaminated environment and food (even if these groups have the technological means to monitor radiation) requires a constant, unceasing effort that is often well beyond their resources. The chronic character of contamination means that infrastructural solutions also have to be sustained over time. And the greater the recognized scope of danger, the greater the scope—and potentially complexity and cost—of the required infrastructural solutions. The government's own interest in reducing the costs of mitigating contamination may coincide with pressure from industry to lower protection standards.[4] Just as banks have been described as "too big to fail," some hazards might be too big to be recognized and mitigated, especially if their recognition would require mass evacuation or compensation for the affected groups. (The imperceptibility of the hazard and the delayed nature of its effects might also make it more difficult to demonstrate the beneficial effects of protective measures when they are actually attempted; it might not be clear what imminent harm has been avoided.)

Even wealthier nations can easily become overwhelmed by chronic and pervasive hazards. Three months after the Fukushima nuclear accident on March 11, 2011, one Belarusian radiologist described to me a sense of déjà vu from observing the tactics of the Japanese officials who downplayed the scope of the accident, withheld data, and raised the limits for radiation exposure at least 20 times, including for children.[5] Days after the accident, the Japanese government moved to evacuate the residents, about 130,000 people, of a zone of 20 kilometers (12.4 miles) around the plant. The U.S. Nuclear Regulatory Commission advised Americans within 50 miles of the plant to leave. Applying the same standard to Japanese citizens would have required the evacuation of 2 million people. Yet compared to Belarus, Japan has better technoscientific capabilities and a different civil society with different capabilities for articulating the hazard (as illustrated by the work of civil organizations such as Safecast, which produced and distributed radiation meters to the general population).[6] Fukushima has even presented market opportunities: private companies offered to test consumer food products for a fee and provide food delivery services that routinely tested for radiation.[7] In contrast, a Belrad employee told me that its testing services had to be free for the affected populations—asking them to pay for testing would be immoral. Not everybody would agree with this assertion; nevertheless, market opportunities for post-Chernobyl radiological testing have not been developed in Belarus.

Perhaps more important than commercial opportunities is the role of civil society. In the U.S. context, for example, the visibility of some hazards and environmental illnesses has largely been an achievement of social movements.[8] Environmental movements are missing as a significant collective force in Belarus. The government under President Alexandr Lukashenko has also used various bureaucratic requirements to reduce the number of NGOs, including Chernobyl-related NGOs, and to control the conditions of their work, including, for example, what populations could receive what types of international humanitarian assistance. The suppression of civil society under Lukashenko has no doubt limited the opportunities for articulation of radiological hazards. Yet even under such conditions, civil organizations have remained crucial to maintaining the visibility of Chernobyl, both in Belarus and abroad. Belrad is a notable example. The NGO Belarusian Committee "Children of Chernobyl" is another; it organized annual academic conferences on Chernobyl and published the proceedings, thus maintaining one of the very few sources on Chernobyl research in Belarus.

Analysis of the production of invisibility suggests a number of tactics might be used to promote greater awareness of particular imperceptible hazards. We have seen that creating and maintaining public visibility depends on opportunities and spaces for articulation; even those who are affected by a hazard on a daily basis require opportunities and spaces to articulate their experiences. Visibility also critically depends on the presence of independent experts, not least because of the need to constantly and critically examine the standards of protection. Moreover, democratizing access to the means of monitoring invisible hazards is one of the most important steps for making those hazards publicly visible (and expanding the scope of the considered data).

The public visibility of an imperceptible hazard might also be facilitated by dramatic, even theatrical, public demonstrations of that hazard, in the spirit of Kevin DeLuca's "image events."[9] The picture might indeed be worth a thousand words, especially when it renders a complex phenomenon easily perceivable. Making the invisible visible in this case might be an elaborate work of art, such as when French artists Helen Evans and Heiki Hansen used a high-powered laser to illuminate the clouds of smoke from a coal-burning electrical generating plant in Helsinki in 2008. The laser not only made visible the problem of environmental contamination but also reflected the data on people's energy consumption. The less electricity the local residents consumed, the bigger the neon sign of the green laser got. (In other words, the laser made visible both the cloud and the invisible digital infrastructure that measured the local electricity consumption.) The production of this demonstration depended on the participation of many social actors, including technical experts, a commercial power plant, city officials, and environmental and design groups. The demonstration was clearly empirically based, and it attempted to dramatize what was already there but outside people's attention. Nevertheless, even such visual demonstration still requires, as Evans put it, "a temporary suspension of disbelief" when "reading information into a dynamic cloud."[10] The goal of such image events would be to open up particular hazard-related issues for scrutiny and judgment, even if the visibility constructed in this way is necessarily temporary.

The power of such dramatizations might be in their ability to induce public discussion. Narrative accounts and stories matter no less than scientific information and images in establishing and maintaining public visibility for imperceptible hazards. Those in a position to influence popular accounts—including historians, social scientists, and any public intellectuals conducting research on environmental hazards—thus face particular challenges and even responsibilities. Scholars "might not be able to prevent the powerful from willfully manipulating historical narratives to stay in power," Nancy Langston notes, but "we can and should provide counternarratives that push back against these manipulations."[11]

The production of visibility is critically dependent on an analysis of the power relations. Extreme power imbalances have thwarted our understanding of the health effects of radiation. It is not necessarily wrong to render the hazard less publicly visible. (It is entirely possible that some risks are not as great as they are made out to be. Such assessment might be relative to other risks: vaccines could be hazardous, but the risk of not vaccinating could be so much greater.) Yet without attention to power, any social-scientific or historical consideration of the hazard potentially reaffirms the existing power imbalance. Such analysis naturalizes the estimates of harm produced by those in power and conceals the work done to make the hazard publicly invisible. Without attention to the disparities of power, social-scientific and historical accounts (including those produced by local scholars) can just as easily interpret the greater recognition of the post-Chernobyl consequences in the last years of the Soviet Union as nothing but an information boom, perhaps even, as the IAEA reports suggest, increasing people's anxiety and causing adverse health effects.[12] Such accounts are possible and even plausible since the visibility of imperceptible hazards is always obviously constructed. The question to consider then is what social mechanisms guarantee that our knowledge about imperceptible hazards is adequate. Paying closer attention to the social mechanisms of knowledge production and how they are shaped by power relations may bring us one step closer to environmental knowledge that strives to be socially just.

Appendix: Data and Methodology

This research has relied on a wide range of sources, including interviews, documents, and media coverage. My general methodological approach has been guided by grounded theory methodology.[1] This approach emphasizes analytical induction and the ongoing analysis of data. Data analysis starts from the earliest stages of research. Data collection is directed and controlled by the emerging theory: concepts and hypotheses emerge from the data and point to the next steps and sources. The criterion for when to stop this ongoing sampling is the "theoretical saturation" of the emerged categories, defined by Barney Glaser and Anselm Strauss as the point at which "no additional data are being found whereby the sociologist can develop properties of the category. As he [or she] sees similar instances over and over again, the researcher becomes empirically confident that a category is saturated."[2] To achieve saturation of the categories, the researcher seeks to uncover inconsistencies within the data and collect further data on cases that could potentially contradict the emergent categories.

Guided by this general methodological approach, I followed several lines of data collection. First, I conducted interviews with those working on Chernobyl-related issues in Belarus: scientists and physicians, government administrators, members of NGOs, and members of international projects. I also conducted interviews with several evacuees and cleanup workers in Minsk, but then I focused on interviewing laypeople living in the contaminated territories. Most of these interviews were conducted during trips with radiologists from the Institute of Radiation Protection "Belrad" and with the international program CORE. (The field trips and interviews are described in more detail below.)

Second, I attempted to historically reconstruct transformations in the visibility of Chernobyl in Belarus. The importance of understanding the history of these transformations became apparent to me as I was conducting interviews for this project; it became clear that my interviewees were consistently responding to and commenting on some extended history of the public discourse on Chernobyl. To capture historical transformations of this discourse, I conducted content analysis of 20 years of Chernobyl coverage in four Belarusian newspapers (see below). The following documents were also collected and analyzed as primary sources: national and international reports on Chernobyl, radiation protection booklets, texts of Chernobyl laws and related regulations, Belarusian scientific publications and journals, and Internet sites of key relevant organizations. Many of the document sources collected (such as national reports on Chernobyl) have limited distribution, and I relied on my interviewees to learn about these sources and gain access to them. Only chapter 3 describes Chernobyl media coverage directly, but the observation that emerged from this analysis—that the visibility of Chernobyl fluctuated historically—has informed my argument in all the chapters. Indeed, a systematic analysis of extended periods of media coverage might be the only way to trace the transformations of Chernobyl's consequences in the public discourse. An awareness of the extent and historical timing of these transformations later informed my analysis of the interviews and other data.

Analysis of Chernobyl Media Coverage

For the media analysis I chose to focus on four newspapers:

1. *Sovetskaya Byelorussiya* (currently published under the title *Belarus Segodnya/ Sovetskaya Byelorussiaya* [Belarus Today/Soviet Byelorussiya]) is the main official newspaper, published in Russian. In order to provide the most comprehensive and accurate analysis, I used exhaustive sampling for 1986–2004. I accessed hard copies of the newspaper for 1986–2000; *Belarus Today/ Soviet Byelorussiya* articles for 2001–2005 were accessed through online newspaper archives. In total, I found and analyzed 550 Chernobyl-related articles of various lengths, from a couple of paragraphs to several pages.

2. *Gomelskaya Pravda* [Gomel Pravda] is the main local newspaper in the most affected Gomel region. In 1986 it was published in Belarusian and was issued five times a week. Starting in 1996, it appeared four times a week, then dropped to three times a week in 2000. The language of the newspaper changed to Russian in 1998. I sampled six months every other year, starting from 1986 and including 2004. This sampling fit with the annual pattern of Chernobyl-related coverage: more articles appeared around the date of

the accident (April 26) and at the end of summer, the main agricultural season and the time for gathering the mushrooms that are known to accumulate particularly high concentrations of radionuclides. Chernobyl-related articles during other months appeared less frequently. The total number of *Gomelskaya Pravda* articles included in the analysis was 349.

The National Library, where I accessed the newspapers, was missing a number of 1992 issues of *Gomelskaya Pravda* (including all May and June issues), but it did include issues for April, July, and most of the rest of the year. To compensate for the missing months, I also studied the issues published from October through December of that year.

3. *Ekologicheski Vestnik* [Ecological Bulletin] was a monthly supplement to *Gomelskaya Pravda* beginning in 1990 and became a national paper in 1993. Among the cofounders of this newspaper was the State Committee on the Consequences of the Catastrophe at the Chernobyl Nuclear Power Plant (or the State Committee on Chernobyl). The *Ekologicheski Vestnik* sample included all four available 1990 issues of as a *Gomelskaya Pravda* supplement; all April, May, August, and September issues of 1993; and all April and the first week of May issues from 1994, 1996, 1998, and 2000. I selected April and the first week of May issues to capture the anniversary coverage in *Ekologicheski Vestnik*, and I chose the August and September issues to reflect the agricultural and mushroom season. In total, I analyzed 226 articles from this newspaper.

4. *Narodnaya Volya* [People's Will], the most prominent independent newspaper in Belarus, was established in 1995 by Iosif Seredich.[3] It began as a weekly newspaper, published in Russian and Belarusian. The first issue was July 11–17, 1995. Since 2000, it has been published daily Tuesday through Saturday. My sample included all issues from April through September of the following years: 1995, 1996, 1998, 2000, 2002, and 2004. In total, I found and analyzed 56 Chernobyl-related articles from this newspaper.

I accessed back issues of the four selected newspapers at the National Library in Minsk. All selected articles were photocopied. During the later stages of analysis, I excluded stories that under closer inspection appeared not to be explicitly related to the Chernobyl accident and its consequences. I selected articles based on one criterion: explicit reference to the accident and its consequences. Articles that did not state the connection to Chernobyl explicitly (even when they could have been prompted by the accident, its anniversary, or other closely related events) were not included. For example, a number of articles in 1986 reported nuclear accidents or near-accidents at power stations in Western countries (to demonstrate, one might assume in the logic of the Cold War, that nuclear accidents are normal and that other governments cover them up); such articles were not included. Similarly, if a later article described the rise of child cancers in the contaminated regions but did not make the connection to the accident, it was not included in the sample. Choosing articles based on these criteria was relatively straightforward for the first 10 years of coverage. In the later years, Chernobyl was often mentioned in passing, and Chernobyl-related references (e.g., "Chernobyl benefits" or "Chernobyl cleanup workers") could appear in practically any article. My focus remained on the articles that addressed the accident and its consequences directly.

Articles in Belarusian newspapers cannot always be separated into such categories as news, opinion pieces, and letters to the editor. All the articles (regardless of whether they could be interpreted as news or commentary) were treated equally. To account for the relative significance of the articles, I recorded their size, headline, and the pages that they appeared on. (The pages of a newspaper played particular roles in the Soviet period: official news appeared on the first page of *Sovetskaya Byelorussiya*, whereas TV programs and entertainment pieces appeared on page 4. Pages 2 and 3 contained news from the regions, propaganda items, and reports.)

I conducted quantitative content analysis and qualitative analysis of the media stories. Using content analysis, I arrived at the major themes of coverage during different periods. A more nuanced description of the framing of the coverage—including the scope of Chernobyl's consequences assumed within various frames—was based on qualitative analysis. To select codes and analyze articles, I followed the same procedures for all four newspapers. The unit of analysis was an article. I coded for recurrent themes; all identifiable themes were coded once per article. I arrived at an extensive list of more than 100 themes, which were then aggregated into broader topics and then several groups of topics.[4] The particular transformations of discourse and the range of attitudes were captured through qualitative analysis.

Qualitative analysis of the coverage was done for each newspaper, with particular attention to *Sovetskaya Byelorussiaya*. All photocopied articles were divided into six periods (corresponding to changes identified by content analysis), and each period was analyzed separately. The qualitative analysis was also guided by dominant themes and topics identified for each period during the content analysis. The goal of the qualitative analysis was thus to enrich the codes that emerged from the quantitative analysis of the data and to contextualize and historicize their description.

Interviews and Ethnographic Observations

The interview sampling aimed to include representatives of different expert and lay positions to uncover a range of perspectives on Chernobyl's consequences. It included experts with different institutional affiliations and areas of expertise, and experts and laypeople living in different parts of the country ("clean" versus contaminated areas). It became apparent early in the research that what groups one might identify as relevant to the Chernobyl problem depended on how the problem itself was defined. The boundaries around different groups affected by the Chernobyl accident or its fallout—including, for example, liquidators and the affected populations—were redrawn with each reassessment or reinterpretation of Chernobyl's consequences and frequently remained a matter of social tension. I approached the problem of identifying relevant groups somewhat differently when selecting lay and expert interviewees (see below). All interviews were transcribed, coded, and analyzed using the grounded theory methodological approach mentioned above.

It was often difficult to separate experts and laypeople. For example, some experts were part of the affected population themselves. Scientists or physicians who worked in organizational contexts potentially related to Chernobyl (e.g., the Ministry of Health, the Republican Center for Oncology and Hematology, local hospitals, and departments of radiobiology or radioecology) often claimed the lack of a particular expertise and preferred to express their personal understanding. Therefore, the division into lay residents and expert categories below is tentative; it refers mostly to the interviewees' institutional affiliation and is used only for the purposes of description in this appendix.

The category of experts here refers, rather loosely, to individuals whose professional activities are related to Chernobyl knowledge production practices. It includes scientists, government authorities (including representatives of the State Committee on Chernobyl and the Ministry of Health), physicians, members of international organizations and programs, and selected members of civic Chernobyl organizations. I interviewed about 50 experts, including those living in some of the contaminated areas. Some were interviewed repeatedly during 2004–2012.

My selection of experts to interview relied on "snowballing" sampling. Some experts were approached at a Chernobyl-related conference in Minsk. Interviews with experts tended to last longer than interviews with the affected populations—up to three hours, with several follow-up sessions. Most expert interviews were one hour long (in a few cases, however, the interviewees refused to talk for longer than 10–15 minutes). The interviews were semistructured and depended on the experts' areas of specialization. Compared to interviews with laypeople, these interviews were generally easier to sustain: experts, whose professional activities were related to Chernobyl, were significantly more articulate on the topic. Perhaps one of the key differences between expert and lay interviews was the impression that experts could see the point behind the conversation: the questions I was asking were meaningful to them in a practical way; in some cases, these individuals were passionate about the topic. This contrasted with most lay attitudes. Lay residents of either "clean" or contaminated areas often appeared to be surprised by the topic of Chernobyl and appeared uninterested in or even resistant to talking about it.

Interviews with experts presented a different challenge. The interviews with residents were completely anonymous. In most cases, I did not know or ask for their last names and was only visiting their areas temporarily. With experts, I often sought interviews precisely because of their names and expertise. In some cases, their expertise or position was unique, which obviously complicated the issue of guaranteeing anonymity, particularly in a small country such as Belarus. These considerations led to some narrowing of my list of interviewees as well as the scope of issues addressed in particular interviews. Most expert interviews were tape-recorded; in some cases, I took handwritten notes, depending on the context and interviewees' preferences.

The interviews typically started with a discussion of the history of the experts' professional involvement with their work, their understanding of Chernobyl's consequences, and the relevant activities of their organizations. In the second part of the interview, I often presented my interviewees with opposing viewpoints (e.g., of the international nuclear experts or local administrators). This dialogical context often provided an opportunity for clearer, more explicit reflections. Some of the interviewees also volunteered descriptions of the history of relevant interactions. Depending on the experts' positions, I also sought their opinions on the informing and engaging of the broader public. In addition, I noted any signs of the experts' personal concern (or lack of concern) with radiation danger and past, present, or future consequences of Chernobyl.

Uncovering the perspectives of various lay groups on the topic of Chernobyl radiation danger raised certain methodological challenges. Hardly any activities in everyday life related to radiation safety, and even in cases when connections could be made, it was not always clear whether they were made by the laypeople themselves. In my preliminary trips to assess the scope of possible study, it appeared unlikely that I would be able to learn about laypeople's understanding of radiation danger by following them in their daily lives. (This does not mean that laypeople's understanding of radiation danger was not reflected in their daily activities, only that it was not made transparent in everyday discussions and practices.) Furthermore, I was faced with extensive heterogeneity within Chernobyl-affected areas, including different levels of contamination, histories of resettlement, and socioeconomic and other conditions.

Consequently, my data collection relied on interviews and observations conducted in the contexts in which activities were explicitly related to Chernobyl, such as radiological assessment and meetings organized by Chernobyl projects. Most of the interviews with lay residents of the contaminated areas were conducted during trips with radiologists from Belrad and a team from the international CORE program. Radiologists from Belrad traveled to contaminated rural areas to measure internal radiation doses (i.e., exposure from radionuclides consumed with food products) in schoolchildren. About 100 children were tested on that trip. Trips with the CORE team were to other and more significantly contaminated areas. The team was collecting local initiatives for its socioeconomic rehabilitation projects and held introductory meetings in three villages; about 60 residents attended.

During these trips, I interviewed 37 local residents. The majority, though not all, of the interviewees were people who participated in the activities of the organizations I came with, and most of the interviews were 20–30 minutes long. Interviews conducted during the Belrad trip included several schoolchildren and staff members who were supervising children through the process of measuring and who were getting tested themselves.

Five interviews with people who had been resettled from the most contaminated areas in the Gomel region were longer (up to an hour), and the interviews were conducted in Minsk. Residents of the contaminated areas were less interested in talking about radiation, especially outside the contexts of radiological assessments or Chernobyl-related projects (see discussion in chapter 1).

The lay interviews typically included questions about the past and present scope of radioactive contamination in the area, individuals' own history of radiation-related activities and concerns (e.g., radiation protection measures, use of dosimeters, specialized farming techniques, limitations on forest use, use of private plots), sources of information (e.g., mass media), general understanding of radiation danger and its health effects, local experiences of the Chernobyl consequences (including its economic effects), and radiation-related education of children.

For comparison, I also conducted interviews with lay residents of comparable social groups living in Minsk as well as several cleanup workers and evacuees residing in Minsk. In the final text, these interviews, as well as several interviews with cleanup workers and members of local Chernobyl-related NGOs, appear only as sporadic illustrations. Activities of Chernobyl-related NGOs, in particular, was an area of investigation too complex to fit into the present scope of study—although the history of local Chernobyl organizations would also have developed in waves similar to the waves of (in)visibility described in chapters 3 and 6. In addition to the interviews and observations described above, I visited the Laboratory for Food Irradiation Detection, the Center for Hygiene and Epidemiology (Minsk), food markets in Minsk and in district centers in the Brest and Gomel regions, and the State Committee on Chernobyl.

Document Analysis

International organizations' reports on Chernobyl—such as those by WHO, the UNDP, UNICEF, UNSCEAR, IAEA, the World Bank, and the Chernobyl Forum—are freely available online. (Selected UNICEF reports on the organization's activities in Belarus were also provided by a member of the UNDP staff in Belarus.) General information booklets on the international programs, such as ICRIN and CORE, were obtained from the CORE office and the State Committee on Chernobyl. Belarusian national reports on Chernobyl's consequences, and relevant legal documents were obtained from expert sources and the State Committee on Chernobyl staff and were subsequently analyzed as primary data.

Other documents informing the analysis in this study include the following: annual proceedings from 1996 to 2005 of the conference Ecological, Medical-Biological, and Socioeconomic Consequences of the Catastrophe at the Chernobyl Nuclear Power Plant; research collections published by the Center for Radiation Medicine; radiation protection booklets or publications by civil organizations; and local publications and proceedings of conferences on the psychological effects of Chernobyl. Additional data included newspaper articles given to me by interviewees. I am particularly grateful for the copies of drafts of the revised Belarusian concept of radiation protection provided to me by Astrid Sahn, as well as her selection of newspaper articles on this topic.

The study also utilized numerous miscellaneous sources—most often obtained from interviewees—as primary and secondary data. These sources included written personal reflections or historiographic descriptions of the Chernobyl aftermath by Belarusian experts, translated volumes of Western nuclear critiques, materials from local institutes or the Belarusian-American thyroid project, documentaries (such as *Chernobyl Heart* and *Nuclear Controversies*), and selected articles from the Western media.

Notes

Preface

1. UNSCEAR (2000). Based on these and similar reports and press releases, Polish journalists Rotkiewicz, Suchar, and Kaminski (2001) referred to Chernobyl as "the biggest bluff of the 20th century," and the *Economist* reported that there is "little to fear but fear itself" (September 8, 2005). Similar headlines appeared in other Western media as well (see chapter 5).

2. Residents of the contaminated areas are exposed to increased levels of radiation not only externally; they also consume radionuclides with their food. Estimating Chernobyl's radiation health effects is a matter of assessing the effects of this chronic, so-called low-dose exposure—that is, exposure that does not cause acute radiation sickness. At the same time, incidences of acute radiation sickness might also have been severely underreported by the Soviet government (see chapter 4).

3. Paine (2002).

Introduction

1. This is compared to 5 percent of Ukraine. Shevchouk and Gourachevsky (2001).

2. Belarus Committee on Chernobyl and UNDP (2005).

3. All of our experiences are mediated: we try to make sense of, interpret, what is happening to us, which always requires language.

4. I use the term *Chernobyl's consequences* as the English equivalent of the Russian *chernobyl'skie posledstviya*, a phrase used in Belarus to refer to the aftermath of the accident and the consequences of the Chernobyl fallout.

5. Proctor (1995); Proctor and Schiebinger (2008).

6. Latour (1988).

7. Beck (1992), 71. See also Beck (1995a, 1995b).

8. Beck (1995b), 184.

9. Ibid., 125, 62.

10. Shevchouk and Gourachevsky (2003), 4. This national report, an official publication summarizing the consequences of Chernobyl, starts with the following statement that highlights the nonradiological effects:

Not everybody can imagine the real scope of the tragedy that Belarus has been experiencing in connection ... [with] the explosion of a nuclear reactor at the Chernobyl Nuclear Power Plant in the bordering Ukraine. A significant number of destructive ecological, health, social and economic effects make it impossible to see the consequences of the nuclear accident narrowly from the perspective of radiation safety. One should also keep in mind that the relative weight of the negative consequences was much greater for Belarus than for other affected countries. Subsequently, the Chernobyl consequences in Belarus are more adequately described by such terms as "catastrophe" and "national ecological calamity."

11. Petryna (2002) describes the spiraling effects of disaster remediation in Ukraine as people were trying to fit into the categories of populations and diseases that were compensated. Petryna (2002), 86, also notes that, by 1996, new amendments to Ukrainian Chernobyl laws stopped some resettlement and cut benefits for the inhabitants of the lesser contaminated areas.

12. Petryna (2009), 37.

13. Cresnon (1979); Langston (2011); Michaels (2008); Murphy (2006); Oreskes and Conway (2010); Proctor (1995); Proctor and Schiebinger (2008).

14. Hess (2007); McGarity and Wagner (2010); Oreskes and Conway (2010).

15. Researchers in science and technology studies have often emphasized articulation as a material process—see Latour (2004); Mol (2002); Murphy (2006); and Soneryd (2007). Murphy (2006), 183, in particular, emphasizes material constructions and "materializations" of imperceptible phenomena such as the complex compositions of chemicals. She notes, "The verb *articulate* is useful because it refers not only to speech but also to physicality, such as the way the joint articulates an arm."

16. Dialogue is more than a reference of one perspective (or one text) to another. It is a principal form of the coexistence and interaction of two or more different discursive perspectives on the same problem—see Bakhtin (1981, 1984) and Kuchinsky (1988). Bakhtin (1981), 342, juxtaposes dialogue to monologue, in which there is only one perspective "binding" its audience from the position of authority (e.g., the "authoritative word" of religions, teachers, or parents). Unlike monologue, dialogue includes at least two perspectives, which means that the same issue or object is presented in at least two different ways: *as something* and as *something else*, according to Kuchinsky (1988).

17. Consequently, mass media in different countries have their own dialogues on Chernobyl.

18. Bakhtin (1981, 1984) refers to different perspectives in a dialogue as *voices*. A voice always reflects a particular meaningful—or, more precisely, meaning-generating—position; thus, the understanding of one's perspective allows for a degree of anticipation of what is going to be said. According to Kuchinsky (1988), each utterance in a dialogue is a response not just to the previous statement but also to the whole narrative as it has been jointly constructed thus far. If each statement reflects and builds upon the whole body of the narrative created by both positions thus far, then the two positions are in principle shaping each other, yet the unfolding of the dialogue is the outcome of there being two separate, distinct meaningful positions (which neither fully accept nor reject each other).

19. My approach to defining infrastructures builds on Bowker and Star (1999) and Star and Ruhleder (1996).

20. It might be more accurate to say that particular aspects of radiation health effects become unknowable within a given context, but some contexts—like the post-Chernobyl exposure of large numbers of people—are unique and cannot be ethically reproduced.

21. Bowker (2005a); Latour (1987).

22. Malko (1998b).

23. Connecting the rise in adult thyroid cancers to radiation exposure has proved to be significantly more controversial, even for the population of the most affected regions.

24. Radiation-induced conditions may include a range of cancers (including adult thyroid cancer and breast cancer) and an wide range of noncancerous morbidity (including heart disease).

25. Star (2006), 1.

26. Bowker and Star (1999) have described how phenomena can be made invisible from perspectives embedded in particular technological infrastructures. In communication studies, Gitlin (1980) has described the making and unmaking of the New Left in the media. Consider also the discussion of invisibility by Hecht (2012) and Greene (1999).

27. Murphy (2006); Proctor and Schiebinger (2008).

28. Proctor and Schiebinger (2008), 22.

29. The 2001 map was available for purchase from bookstores in Minsk in 2003 when I first attempted to find a map of the contaminated areas in Belarus. Another edition appeared in 2004, but it was not as easy to find. See chapter 4.

Chapter 1

1. Similarly, Gould (1993) argues that the intuitive assumption that people living closer to pollution sites are more likely to organize and attempt to reduce their exposure is not always accurate.

2. Alexievich (1999).

3. Beck (1992), 27.

4. Wynne (1992); see also Paine (1992), Wynne (2003, 2008).

5. Scott (1992), 23, reminds us that to simply accept experience as a source of knowledge assumes an unproblematic connection between visibility and knowledge—that "vision is a direct, unmediated apprehension of a world of transparent objects." Scott, a feminist historian, argues against this approach that "takes meaning as transparent" instead of examining how it is reproduced through particular ideological systems in which social categories, such as gender and race, are naturalized.

6. Alexievich (1999), 20.

7. Fowlkes and Miller (1987), 55–56.

8. Latour (2004), 225. As Soneryd (2007), 288, points out, another advantage of this approach is that it calls into question the search for accurate representations of reality, or "an accurate, stable referent." Positions of experts and laypeople thus become more dynamic and changing.

9. In her discussion of sick building syndrome and related consciousness-raising, Murphy (2006) reminds us that the tools and methods used in experiencing are always time- and place-specific—in some cases allowing experience to obtain the status of knowledge.

10. Belrad was established in 1990 by Vassily Nesterenko with active support from the renowned nuclear physicist and dissident Andrei Dmitrievich Sakharov, world chess champion Anatoly Karpov, and Belarusian writer Ales Adamovich. Most of Belrad's funding comes from NGOs abroad. Belrad provides a description of its activities on its website, http://belrad-institute.org. Activities of Belrad will also be discussed in chapters 2 and 4; CORE will be discussed in chapter 5.

11. See Paine (2002) for a description of some cultural contexts in which an effort is made *not* to construct risks.

12. Goodwin (1994).

13. See, e.g., Miller (1994).

14. "Clean" (i.e., uncontaminated) and "contaminated" spots were identified based on the facts of resettlement and through maps of radioactive contamination of the district that used to be displayed in the radiation control center (since closed).

15. In some cases, the baseline might not even be "clean," but there is almost always a general awareness of the differences between places.

16. The term *practical radiological culture* assumes that residents of the affected districts can and should be responsible for reducing their exposure. We will return to the challenges of this individualized approach in chapter 2. Relative successes and problems of the CORE program are also considered in chapter 5.

17. Brown (1992), 269; Brown (1997, 2007); Kroll-Smith, Brown, and Gunter (2000).

18. Beck (1995a), 13.

Chapter 2

1. My analysis focuses mostly on the perspectives of people currently living in the contaminated territories and not on other groups of affected populations, including resettlers and Chernobyl cleanup workers. I address this narrow focus at the end of the chapter.

2. Some international organizations refer to the lifestyle changes required to minimize one's radiation dose as *practical radiological culture*. Zoya Trofimchik, head of CORE, interview, Minsk, 2004.

3. More than two decades after the accident, individual doses are much lower than they were in the first years after the accident. According to radiologists from Belrad, however, the current doses decrease not as a result of "natural" spontaneous processes such as radiation decay but as a result of organized infrastructural efforts. Belrad personnel, interviews, Minsk, 2010 and 2012.

4. All village names used in this chapter have been changed, but the text preserves the original names of the districts, since the affected districts vary greatly in their social and radiological circumstances.

5. World Bank (2002), 6, 10.

6. Belookaya (2004b, 2004c). The study was conducted as part of the International Chernobyl Research and Information Network project to study "the information needs" of "the affected population." See chapter 5 for further description.

7. Edelstein (1988).

8. Fowlkes and Miller (1987). The role of sociodemographic factors, different risk exposures, and experiences has been noted by other researchers, such as Powell, Dunwoody, Griffin, and Neuwirth (2007).

9. Fowlkes and Miller (1987), 61, 72.

10. Consequently, perspectives expressed two decades after the accident differ from those described by earlier oral histories of Chernobyl, such as Alexievich (1999). The more recent perspectives have assimilated past efforts of "educating the public" about radiation danger, state Chernobyl policies, and popularized scientific perspectives.

11. According to the dialogical approach, tensions between this echoing and what is unique and individual in these expressions are irreducible. Bakhtin (1981, 1984); Kuchinsky (1988); Wertsch (2002).

12. Billig (1998) offers a wonderful example in which individuals (an ordinary British family) *jointly re-create positions* for and against a particular topic (the role of the royal family). Wertsch (2002), building on Mikhail Bakhtin's theory, also shows that narratives are not just a matter of individual construction; they always reflect not just the author's voice and those whom they address but also others who have used these words and made these statements before. The echoing effects of individual statements about Chernobyl were extremely noticeable in my interviews with laypeople and with certain experts. Individual judgments in these cases often had recognizable "tunes" or sentiments.

13. One's actions can, however, be reinterpreted retrospectively and often are.

14. Belrad personnel, interviews, Minsk, 2004, 2010.

15. Belookaya (2004b, 2004c).

16. Most food markets have centers for radiation control. Individual sellers are required to have their produce tested and to be able to show certificates from these tests to the buyers. However, the sellers could be selling both uncontaminated and contaminated food, manipulate which items are tested, or lie to the buyers about which areas their produce is from.

17. According to the official records, the state infrastructure includes 850 units of radiation control, performing 11 million tests for Cesium-137 and 18,000 tests for Strontium-90 annually. Belarus Ministry of Emergencies (2011), 61.

18. Skryabin (1997). According to ASPA (2009), 86, the doses of rural residents are still 1.3–4.0 times higher than those of urban population.

19. Skryabin (1997). Few years after the accident, the residents of these villages went back to consuming milk from their cattle faster than the residents of larger villages did, and the cutback was generally less than 30 percent (compared to 70 to 80 percent in larger localities). The effectiveness of prohibitions on the use of forests was even less noticeable. Skryabin argues that the measures were imposed top-down and went against the local way of life and the people's customary reliance on forests. He fails to account for severe economic shortages in the last years of the Soviet Union,

when the prohibitions were imposed, as well as the general lack of adequate infra-structures (including grocery shops) in remote rural communities.

20. Belrad personnel, interviews, Minsk, 2004. Belrad radiologists used the Chech-ersk district as an example of another particularly difficult district that still has vil-lages that are relatively well-off.

21. Belookaya (2004b, 2004c).

22. Ibid. *Curie* is a unit measuring the intensity of emitted radiation, and the number of curies per square kilometer describes the contamination of a site.

23. Skryabin (1997). The sociological advantage of describing individual risk behav-iors in the context of post-Chernobyl radiological contamination is that the differ-ences can be assessed rather precisely by measuring people's internal radiation doses (i.e., exposure from radionuclides that have been consumed with food).

24. Powell et al. (2007) summarize the results of previous studies (for Western con-texts), arguing that women are more likely to be concerned about health risks.

25. Skryabin (1997) interprets the difference between these groups as a difference in psychological and lifestyle characteristics. Much like the Soviet nuclear experts (or international reports described in chapter 5), Skryabin argues that high levels of internal radiation are a matter of bad individual lifestyle choices. He also considers doses to be generally too low to cause health problems. His views were reported in *Gomel'skaya Pravda* on April 25, July 16, and November 19, 1992, and on April 25 and 26, 1996. According to Skryabin, anxiety related to unnecessary concern with radiation is ultimately more harmful for individuals than radiation exposure. His estimates show the level of anxiety to be three to four times higher for the small-dose group than for the high-dose group. Skryabin (who worked at the Soviet nuclear sites and came to Belarus in 1986 at the invitation of the minister of health care) was also one of the supporters of the Soviet Safe Living Concept of radiation protection (see chapter 4). Approaches similar to Skryabin's, emphasizing anxiety and lifestyle as more significant problems than post-Chernobyl radiation exposure, will be discussed in chapter 5.

26. Belrad personnel, interviews, Minsk, 2004.

27. Belrad personnel, interviews, Minsk, 2004, 2005. The acceptable annual dose in Belarus is 1 millisievert per year. The dose is not measured directly but estimated (see chapter 4). According to the calculations of the Belarusian Ministry of Health, the annual dose of 1 millisievert corresponds to radioactivity of 361–433 becquerels per kilogram, depending on age. Babenko (2003), 39.

28. Regardless of what percentage of the overall dose is attributed to internal expo-sure, Belarusian experts typically agree that internal exposure is more dangerous and that the overall doses two decades after the accident are significantly lower than they used to be, even in the early 1990s.

29. Health recuperation—a temporary residence or vacation in uncontaminated territories, with the corresponding consumption of food that is not radioactively contaminated—was first introduced as a prevention measure in the 1989 Chernobyl Program (see chapters 3). The goal was to decrease exposure doses for the populations residing in the contaminated territories. Its adoption as a public health measure was related to the practical impossibility of providing the residents of the contaminated areas with uncontaminated food. (Critics emphasize that the doses are reaccumulated soon after children, or adults, return to their places of residence and their usual diets). In the 1990s, health recuperation programs for "children of Chernobyl" from Belarus were also organized by various international humanitarian organizations and nongovernmental organizations. Their efforts were hindered by the Belarusian government in the late 1990s and even more significantly in 2000 and thereafter. In 2005, the Belarusian government expressed concern about the ideological influence of the health recuperation trips abroad for children from the affected territories and suggested that international humanitarian organizations instead sponsor children's rehabilitation within Belarus. At the same time, government-funded health recuperation for schoolchildren during the school year (when children travel to recreational facilities in other parts of the country) is often viewed in a negative light as "forced." Belookaya (2004b, 2004c).

30. Aglaya's center is now supervised by the Brest branch of the State Institute of Medicine.

31. Aglaya's village differs from other villages in the extent of attention it received from international projects. But any specific locality has a unique constellation of factors, and Aglaya's description of her village is a good starting point for thinking about other communities.

32. Indeed, Belrad radiologists themselves say that it is not enough to teach people radiation-minimizing techniques, and they advocate the use of Vitapekt, an apple-based absorbent that they distribute after conducting WBC testing. Both Vitapekt and WBC testing are free of charge. In the words of one radiologist, "to *sell* it in the contaminated areas would be atrocious." Interview, Minsk, 2012.

33. The community also benefited from the attention of two international projects.

34. Aglaya acknowledged that she would need help from experts converting the radiation doses in food products into background activity in the area, which in itself shows her command of the relevant scientific knowledge. "In my mind," she said, "I see it all in different colors; each area would have its own number."

35. There is a paradox here. According to Belookaya (2004b, 2004c), young people are typically more knowledgeable than other groups about radiation danger. Nevertheless, they often did not want, and still do not want, to stay in the contaminated territories. This trend leads to skewed demographics and lower birthrates in some

areas. Belookaya reported that 73 percent of the participants in her survey, all residents of the affected territories, did not want their children to stay in these regions, which they associated with radiation and a low quality of life ("Fate has it that we live here, but our children do not have to"). Residents often lack basic social security, including employment, stable future prospects, and adequate health care. As a result, some areas of evacuation and resettlement are populated mostly, or even exclusively, with elderly retirees who did not want to leave their homes or who have since returned there.

36. Zoya Trofimchik, head of CORE, interview, Minsk, 2004.

37. Belookaya (2004b, 2004c).

38. The Forgotten Villages project is discussed in chapter 4. As part of this project, Belrad radiologists conducted WBC testing of internal radiation doses in villages excluded from the official list of the affected localities. This reclassification had led to people losing their Chernobyl-related benefits and compensation.

39. For example, on one of the trips described in chapter 1, the head of the high school made copies of the list of doses in preparation for a meeting with the local administration.

40. Wallman (1998), 170.

41. Aglaya, for example, justified her belief that free lunches should continue at school by stating their general value for children's health and not specifically for Chernobyl-related reasons:

After all, children are not going to have the same kind of food at home. They are not going to have meat every day. Only in winter, when people start butchering pigs, will there be meat for a month, not longer, and then it is only fresh for a day or two, and then it's salted; it's not the same. Free lunches in schools matter. Children are going to have an orange there, a banana. Their parents won't buy that for them; they don't have money. So I think it is important to keep free lunches.

42. Head of Belrad's WBC Laboratory, interview, Minsk, 2004.

43. Article 18 of the law addresses the benefits for citizens disabled as a result of the catastrophe. Article 19 describes the benefits for the Chernobyl cleanup workers. See chapters 3 and 4 for the historical context of the adoption of this law and its subsequent revisions.

44. Tonya was also aware of dissenting perspectives. She noted, "A friend of mine says, 'Who cares about radiation when there are so many children born with syphilis?'" Tonya then resolved the disagreement by referring to underlying political causes: "In either case, there is too much covering up, and people are afraid and don't say things as they are."

Chapter 3

1. Beck (1995a), 64, 67.

2. Gamson and Modigliani (1989); Mazur (1987).

3. Beck (1992).

4. See Cox (2010) for a summary of what this means for media reporting on environmental hazards.

5. *SB*, April 28, 1998. Although the Chernobyl anniversary is April 26, in some years (e.g., 1996), the anniversary coverage started as early as March.

6. *SB*, July 3, 1987; *SB*, April 26, 1987. I often use "elimination of the consequences" rather than "liquidation of the consequences" to avoid the English-language connotation of *liquidation*. In 1986–1987, a number of articles were devoted to past and present nuclear accidents in other countries, suggesting that nuclear accidents were normal elsewhere and that they were frequently not reported in time, thus rebutting foreign criticism of the Soviet secrecy in relation to the Chernobyl accident and offering Soviet counterpropaganda.

7. Marples (1996a), 38; Nikitchenko (1999).

8. See Petryna (2002) for a discussion of opinions of the most prominent Soviet and American experts on the post-Chernobyl circumstances. In *SB*, these experts offered reports limited to what was happening at the site of the accident and what were the immediate consequences of it, such as the fate of the first Chernobyl cleanup workers.

9. *GP*, June 14, 1988. The same article was reprinted in *SB*, August 6, 1988. Similar arguments appear in *GP*, June 2, 1988.

10. Marples (1996a); Zaprudnik (1993). The nationalist movement, the main force toward independence in other Soviet republics, was much weaker in the Byelorussian republic than in, for example, the Baltic republics or in Ukraine. (The issue of the revival of the Belarusian language was, however, rather prominent during this period.)

11. *SB*, February 9 and April 27, 1989.

12. Tamara Belookaya, interview, Minsk, 2004.

13. *SB*, May 21 and July 21, 1989.

14. *SB*, April 16, April 27, and July 21, 1989.

15. *SB*, July 1, 1989.

16. For a more detailed description of the Chernobyl Program, see Marples (1996a), 46.

17. *SB*, December 9, 1989.

18. *SB*, June 24 to July 4, 1990.

19. Marples (1996a).

20. *SB*, July 4 and August 21, 1990. The Soviet scientists were asked to respond by October 1990.

21. *SB*, July 4, July 5, and July 13, 1990.

22. Marples (1996a).

23. *SB*, August 12 and August 15, 1990; *SB*, July 27, August 24, and October 26, 1990.

24. *GP*, April 28, 1990. Even though *GP*'s coverage of Chernobyl was similar to *SB*'s and followed similar transformations, *GP* did pay more attention to the political struggles concerning Chernobyl. For example, it explained Chernobyl decrees in more detail, and it sounded the alarm that the voices of Belarusian scientists might be silenced.

25. Marples (1996a).

26. *SB*, January 31, 1991.

27. *SB*, August 17 and November 7, 1990.

28. *SB*, January 18, 1991.

29. *SB*, March 28, 1991.

30. *SB*, August 2, 1990.

31. *SB*, March 19, 1992.

32. Zaprudnik (1993).

33. Marples (1996a).

34. Ibid.; *SB*, March 30, 1996.

35. Marples (1996a), 64, 66.

36. The situation had many parallels to Petryna's (2002) account of "biological citizenship"—claiming material assistance on the basis of injured health—in post-Chernobyl Ukraine in the 1990s, even though, as we will see below, this period did not last as long in Belarus.

37. *SB*, April 30, 1992; *SB*, August 11, 1992; *SB*, April 9 and May 16, 1992.

38. *SB*, May 21, August 21, and December 30, 1993.

39. *SB*, May 5, 1992. Health generally remained a frequent topic of Chernobyl-related coverage in *SB* at the time.

40. *SB*, February 15, 1992.

41. *SB*, November 19, 1992.

42. On the Belarusian side, local Chernobyl foundations functioned with great disparity—different levels of efficiency and funding; Marples (1996a), 72. Because of the scope of humanitarian efforts from international NGOs, thousands of children traveled abroad, especially to Germany, for health recuperation; *SB*, July 6, 1997.

43. *SB*, April 26, 1996. The survey was carried out by the Belarusian independent service Public Opinion, headed by David Rotman, and it sampled residents from 12 Chernobyl-affected districts in the Gomel, Mogilev, and Brest regions.

44. Ivan Kenik, chairman of the State Committee on Chernobyl, "declared that 'mass relocations' of citizens had been completed." Most decontamination work was completed by 1992–1993. Marples (1996a), 58, 67.

45. *SB*, June 21, 1996. Although Lukashenko was elected in 1994, he began counting his time in office from 1996. A national referendum in 2004 removed the constitutional term limits for presidency.

46. *SB*, July 4, 1995.

47. *SB*, April 10, 1996; *SB*, September 13, 1996.

48. *SB*, March 13 and March 30, 1996; *SB*, March 21, 1996; *SB*, May 27, 1996.

49. *SB*, April 2, 1996.

50. *SB*, November 19, 1997. *SB*, April 22, 1997, stated, "Radiation danger has decreased but hasn't disappeared."

51. *SB*, April 27, 2001; *SB*, April 27, 2004.

52. *SB*, April 26, 2001; *SB*, April 27, 2004, and April 19, 2005. The free economic zones included areas with the right to resettle and areas of secondary resettlement (see the table in chapter 4).

53. *SB*, April 27, 2004.

54. *SB*, April 26, 2005; *SB*, April 25, 2003; *SB*, June 11 and June 20, 1998 .

55. Alexievich (1999), 19. In this collection of oral histories, the section "Interview by the Author with Herself about Missing History" begins with observing the transformation of Chernobyl into a symbol and a thing of the past: "More than ten years have passed. Chernobyl has become a metaphor, a symbol. Even history."

56. *SB*, December 11, 1997.

57. *SB*, December 16, 2000; *SB*, April 26, 2001; *SB*, April 26, 2003.

58. *SB*, January 28, 1999.

59. *SB*, February 20, 1997; *SB*, April 27, 1995. Similarly, representatives of the Belarusian civic organizations who were addressing the president argued that, "Every year 10–15 people injured by Chernobyl pass away before they turn 40. Only 10% of Belarusian children under 14 are practically healthy." *SB*, August 10, 1995.

60. *SB*, September 18, 1998.

61. *SB*, April 21, 2006, http://www.sb.by/post/51066, describes a roundtable with Chernobyl experts organized by SB; *SB*, April 26, 2005, http://www.sb.by/post/43182.

62. *SB*, April 27, 2011, http://www.sb.by/post/115942; *Belta*, September 24, 2013, http://news.belta.by/en/news/society?id=727241.

63. *SB*, September 16, 1997. Critical references appear in *NV*, October 6, 1995.

64. *SB*, October 17, 2012, http://www.sb.by/post/137969; *SB*, April 18, 2011, http://www.sb.by/post/115521.

65. *GP* also included a few more reports on Chernobyl-related science. Several of these stories were by Anatolii Skryabin, whom we met in chapter 2. *GP*, April 25, July 16, and November 19, 1992, and April 25 and 26, 1996. None of these articles challenged the government discourse or drew attention to radiation risks or effects.

66. *GP*, June 30, 1998, and April 25, 2000.

67. The State Committee on the Problems of the Consequences of the Catastrophe at the Chernobyl N[uclear] P[ower] P[lant] was reorganized and renamed six times between 1990 and 2011. In 1995–1997 it even became a ministry: first the Ministry of Emergencies and Protection of Population from the Consequences of the Catastrophe at the Chernobyl NPP, and then just the Ministry of Emergencies. The Committee on Chernobyl reappeared in 1998, although its institutional affiliation changed a few more times. Constant changes made to the Committee reflect the political struggles around the subject of Chernobyl. However, to avoid confusion, I consistently refer to this state organ as "the State Committee on Chernobyl." Institutional changes to the Committee are listed in Belarus Ministry of Emergencies (2011), 34.

68. This is somewhat ironic, as it was predominantly the oppositional intelligentsia, especially the BNF, that raised Chernobyl issues in the last years of the Soviet Union.

69. *NV*, April 25, 1998. Chernobyl Path re-emerged in 1996. Marples (1999), 83.

70. *NV*, April 24, 2002; *NV*, April 26, 2000.

71. *NV*, April 26, 2000, April 26, 2002, April 30, 2002, and April 24, 2003.

72. *NV*, April 30, 2002.

73. *NV*, April 25, 2002.

74. Lampland (2010) describes "fake numbers" used in working documents or business plans. Similarly, the provisional numbers about Chernobyl enable rationalization within these particular local conditions.

Chapter 4

1. A later systematized collection of maps in the 2011 *Atlas of Contemporary and Projected Aspects of the Consequences of the Accident at the Chernobyl Nuclear Power Plant* appears to have a very limited circulation and, to the best of my knowledge, is not available through regular bookstores in Belarus. Its original circulation appeared to be very limited, but *Atlas* was later made available for download from http://www.chernobyl.gov.by.

2. Beck (1992), 46.

3. Star (1995), 90, 92; Latour (1986, 1987); Bowers 1992.

4. Bowker and Star (1999).

5. Ibid.; Lampland (2010); Ottinger (2010).

6. Almklov (2008); Bowker and Star (1999); Busch (2011); Lampland and Star (2009).

7. Star (1995), 101.

8. For description of articulation as work see Corbin and Strauss (1988), Fujimura (1987); Schmidt and Simone (1996); Star and Strauss (1999); Strauss (1985, 1988).

9. Nesterenko (1998), 37.

10. Nikitchenko (1999) provides a firsthand account.

11. Belarus Council of Ministers (2001).

12. Petryna (2002).

13. Ibid.; Marples (1996a); Medvedev (1990); Schmid (2004).

14. New housing, medical facilities, and gas pipelines were built in these areas of resettlement in 1987 and 1988. Several years later, many of these areas were classified as territories of "strict radiation control"; see Matsko and Imanaka (1998) and Yaroshinskaya (1995). Many of the resettled, along with the original residents, had to be resettled again.

15. The phenomenon of unrecorded or understated doses is documented by oral history and witness accounts—see Aleksievich (1999); Lupandin (1998); Petryna

(2002); and Yaroshinskaya (1998a, 1998b). See Imanaka (2008) for a discussion of the acute radiation syndrome cases among the population, thought to be in the hundreds.

16. Matsko and Imanaka (1998); Nikitchenko (1999), 32. Meat with levels of radio-activity exceeding the newly raised acceptable levels was mixed with uncontaminated meat. Contaminated grains were sent to chicken and pig farms, where they were fed to animals, even though it was commonly believed that chicken, eggs, and pork were not radioactive.

17. *Sievert* is a unit for "effective dose" used in radiation protection to estimate the effects of ionizing radiation on a body (taking into account different radiosensitivity of body tissues).

18. Shevchouk and Gourachevsky (2001), 31.

19. Decree reprinted in Nikitchenko (1999), 48.

20. Nesterenko, Nesterenko, and Sudas (2004). As noted in an earlier chapter, the number of curies per square kilometer describes the contamination of a site by measuring the intensity of the emitted radiation.

21. Shevchouk and Gourachevsky (2001), 36.

22. Malko (1998b), 7.

23. Ibid.; Nikitchenko (1999), 145.

24. For example, the concept did not factor in the accumulation of plutonium, strontium, or "hot" particles. According to critics, it did not specifically analyze the irradiation of different groups immediately after the accident, when people lived without any restrictions and without adequate iodine prophylactics, and it did not take into account iodine endemicity (deficiency) in the most affected areas.

25. Nesterenko (1998), 38.

26. *SB*, March 16, 1990.

27. Reprinted in Nikitchenko (1999), 62–65.

28. Malko (1998b). According to Malko, the WHO delegation included Dr. Dan Beninson, the chairman of the International Commission on Radiation Protection (ICRP) and the director of the License Department of the Argentina Atomic Energy Commission; Professor. Pierre Pellerin, the chief of Radiation Protection Services of the French Health Ministry and a member of the ICRP; and Dr. P. J. Waight, a radiation scientist of WHO's Division of Environmental Health.

29. Not all Belarusian experts agreed with the concept, even though it was approved by the Presidium of the Academy of Sciences of the BSSR. In September 1989, 92 scientists, including five Belarusian scientists, signed a letter to Premier Mikhail

Gorbachev that expressed their support for the Safe Living Concept on the grounds that it was supported by the international experts and based on the data from Hiroshima and Nagasaki studies. Critics of the letter, such as Ivan Nikitchenko, described it as "smooth on paper." Nikitchenko (1998), 60.

30. The right to resettle from contaminated areas was critical because of the Soviet system of registered residence (*propiska*), which was required for employment, health care, and most bureaucratic interactions. Some authors of the Belarusian concept went as far as to call for unconditional resettlement "without consideration of the opinions of residents," as well as no agricultural production, in the areas with Cesium-137 contamination of more than 15 curies per square kilometer; *Naviny Belaruskai Akademii*, September 21, 1990.

31. Zgersky (1998).

32. For a discussion of the Chernobyl laws and their implementation, see Marples (1996a).

33. Ibid.; UNDP and UNICEF (2002).

34. Ironically, the Health Ministry itself used levels of soil contamination to estimate external radiation exposure and then average annual doses. The accuracy of the calculations was impossible to prove; Zbarovski, Delin, Malko, Matveenko, Nesterenko, and Rudak (1995). Critics of the concept, such as Vassily Nesterenko and Valeryi Shumilau, also appeared in press; *Belarusskaya Niva*, October 22, 1993, and *Zvyazda*, August 13, 1993.

35. This meant that protective measures would target only 300,000 people out of 2 million previously considered affected, as observed by Alexandr Devoino; *Iskra*, July 21, 1993. Similar observations were made in press by Vladimir Malko and Vassily Nesterenko; October 14, 1993, and *Narodnaya Gazeta*, May 16, 1995.

36. The fact that this concept did not recommend resettlement even in the areas with an average annual dose close to 5 millisieverts per year led Ivan Smolyar, the head of the Supreme Council's Committee on Chernobyl, compare it with Ilyin's Safe Living Concept, which allowed 350 millisieverts per life (estimated at 70 years); *Narodnaya Gazeta*, December 22, 1993. For a discussion of the problems of using averages in risk assessment, see Beck (1992).

37. Critics estimated that people who lived in the areas with 1 curie per square kilometer and who consumed their own food received about 4 millisieverts per year; *Literatura i Mastactva*, July 5, 1994. In smaller rural communities, people are more likely to rely on subsistence farming. According to Nesterenko (1998), these communities were also more difficult to reach. As a result, studies by state and international experts often focused on towns (where people had smaller internal doses), thus creating a false picture of a generally lower range of doses of internal contamination.

38. Mikhail Malko, quoted in *Belaruskaya Niva*, September 29, 1993.

39. *Zvyazda*, July 30, 1996; Nesterenko (1998), 39–40.

40. A common interpretation was that the government was afraid to make dramatic changes to the Chernobyl laws. In 1995, President Lukashenko's decree suspending a number of articles of the Social Protection of Citizens law was canceled by the Constitutional Court. Matsko and Imanaka (1998).

41. Revisions to the concept of radiation protection continued after 1995, but the key principles stayed the same. The 1998 version of the concept reintroduced the levels of soil contamination but kept the same zoning for protective measures (i.e., the area with 1–5 curies per square kilometer received periodic monitoring; territories with 5–15 curies per square kilometer no longer had a "right to resettle" status). The 1998 law on radiation protection specified life doses.

42. Numbers reflecting the list of contaminated communities show some discrepancies; see ASPA (2009), 128, and Belarus Ministry of Emergencies (2011), 10. According to the latter, the list of communities in the zone of radiological contamination was confirmed by decree N132 of the Council of Ministers on February 1, 2010. There are currently 2,402 communities on the list, with more than a million residents.

43. By 2011, the government was considering agricultural rehabilitation of the areas adjacent to the Polesski State Radioecological Reserve and the zone of exclusion, the 30-kilometer (18.6-mile) area around the Chernobyl nuclear plant. *Belta*, April 26, 2011.

44. *Belarusskaya Niva*, October 22, 1993. Nesterenko opposed the Petryaev group's concept and proposed his own alternative. He insisted that protective measures should focus on the most vulnerable groups of the population (i.e., children and pregnant women) and those with the highest doses. To better achieve that goal, he suggested, control levels (0.3 millisievert) should be used in addition to acceptable limits of exposure (1 millisievert per year in access to natural background radiation).

45. Nesterenko et al. (2004); Physician from the former Institute of Radiation Medicine and Endocrinology, interview, Minsk, 2003.

46. Belrad could demonstrate that, for example, the average internal dose of the critical groups (children under age seven and pregnant or nursing women) in the village Olmany in 1992 was 7 millisieverts per year, compared to 3 millisieverts per year stated in the national Catalog of Doses. Nesterenko (1998).

47. Star (1991); Star and Strauss (1999),

48. Belrad personnel, interview, Minsk, 2010.

49. Belrad personnel, interview, Minsk, 2010. A few LCPRs were not closed but were transformed into what were called Centers of Practical Radiological Culture and given

the status of NGOs. As of 2012, six of these centers remained. They functioned primarily as educational facilities (which required permissions from local authorities).

50. See especially Yablokov, Nesterenko, Nesterenko, and Sherman (2009).

51. In addition to testing, Belrad organizes free distribution of pectins and children's health recuperation trips to lower their internal contamination.

52. ASPA (2009), 124.

53. The state's radioprotective efforts have included the evacuation and resettlement of more than 137,700 residents from the Gomel and Mogilev regions; the burial of highly contaminated houses and villages; the decontamination of populated areas; the limitation of access to contaminated areas; the discontinuation of agricultural production and limitations on timber production in the highly contaminated areas; changes to farming practices (plants that tend to accumulate radionuclides were eliminated from local agricultural production); the imposition of limitations on the consumption of contaminated produce; the decontamination of particular farms and production sites; gasification (to limit local use of forest wood for heating); the construction of roads, water supply infrastructures, and housing; food monitoring at more than 1,000 sites of radiation control; the monitoring of soil contamination; radiation-related research (see chapter 6); free school lunches; and health recuperation programs, especially for affected children.

54. Busch (2011), 301.

Chapter 5

1. IAEA, WHO, and UNDP (2005).

2. Tamara Belookaya, interview, Minsk, 2005.

3. Malko (1998b), 14.

4. Although this chapter focuses primarily on the reports issued by IAEA, WHO, and UNSCEAR, their assessments of Chernobyl's health effects are in line with reports by other committees, such as National Research Council (2006).

5. *Washington Post*, September 6, 2005; *New York Times*, September 6, 2005; *Economist*, September 8, 2005; Gofman (1994).

6. Murphy (2006).

7. Charles Briggs, in studying how Venezuelan public health officials and journalists produced information about cholera, argues that top-down, a priori faulty models of the public depend on the actual exclusion of the affected populations. Briggs (2003); Briggs and Mantini-Briggs (2003).

8. Since the focus of this chapter is on the major public reports assessing the Chernobyl radiation health effects, I do not discuss the work of the ICRP, an independent international organization that provides recommendations for nuclear safety standards. It is nevertheless worth mentioning that the standards recommended by the ICRP experts are affected by the dominant approaches to radiation protection and the general disregard for the post-Chernobyl data.

9. Boudia (2007), 390, 400, 402; see also Hecht (2012), 185–186.

10. See, e.g., Yablokov et al. (2009), 325. On April 26, 2007, a number of international civil organizations sponsored the International Day of Symbolic Action and the start of "an indefinite demonstration," a daily vigil in front of the WHO building in Geneva, Switzerland. The organizers were demanding an amendment to the 1959 agreement between WHO and IAEA (WHA 12.40, approved by the 12th World Health Assembly on May 28 1959); see http://www.independentwho.info. In an interview in 2010, Alexey Nesterenko described publication of the May 28, 2009 article "Toxic Link: the WHO and the IAEA" in *The Guardian* as a small victory of the protest. At the time of this writing, the daily vigil in front of the WHO building continued.

11. This structural positioning also appears to favor the IAEA. When asked why the proceedings of the 1995 Chernobyl conference have never been published, Hiroshi Nakajima, the director-general of WHO from 1988 to 1998, replied that it was organized jointly with the IAEA, and "for atomic affairs, ... military use or civic use [of nuclear energy], they [the IAEA] have the authority." Tchertkoff, (2003). Tchertkoff's documentary features the 2001 Kiev conference on Chernobyl's consequences, and it illustrates conflicting expert perspectives on Chernobyl's effects.

12. Fischer (1997), 2, 171. For a decade after the accident, the number of new nuclear power plants built in the West shrunk to almost zero.

13. Stephens (2002), 91–92, 108. Stephens notes that this portrayal is explicitly gendered. She also argues that crucial to these efforts was the boundary between *here* and *there*, implying that "Chernobyl will not happen here in the West."

14. Park (1989), 147. The IAEA experts fully accepted the Soviet explanation of the reasons for the accident itself, but this position was changed in 1993. The Soviet experts attributed the accident's cause to "a violation by the Chernobyl NPP personnel of the procedures of nuclear reactor operation developed in the USSR." Malko (1998b), 5.

15. Petryna (2002), 38, 39. For a discussion of the importance of ignorance in asserting state power, see Mathews (2005). For a discussion of power, ignorance, and knowledge, see Sullivan (1007).

16. Malko (1998b), 5.

17. Nikitchenko (1999); Yablokov et al. (2009), 33. The critics were also concerned with other suspected Soviet practices of rendering radiation danger less visible—for example, by not recording or by misreporting doses, by misdiagnosing affected populations, and by retrospectively altering the data. Medvedev (1990); Yaroshinskaya (1995).

18. Malko (1998b), 8. For a list of the members of the group, see chapter 4, note 28.

19. Quoted in IAEA (1991c), 4–5, and IAEA (1991b), 3.

20. Read (1993), 305.

21. Petryna (2002), 52.

22. IAEA (1991a, 1991b). For the historical examination of the Atomic Bomb Casualty Commission, later renamed the Radiation Effects Research Foundation, see Lindee (1994).

23. IAEA (1991b), 8, 13, 20. The director of RERF, Dr. Itsuzo Shigematsu, chaired the International Chernobyl Project's advisory committee.

24. IAEA (1991a), 32.

25. Petryna (2002), 50.

26. The Belarusian Minister of Health presented these data at the "informal meeting arranged by the IAEA Secretariat on the 19th of December 1989 in Vienna." Malko (1998b), 10.

27. Petryna (2002), 54, cites the example of "hot particles" (radioactive dust and debris from the reactor core) and the necrosed tissue photographs presented by Evgeni Petryaev in the context of the International Chernobyl Project's meeting in Vienna. Hot particles that burned lung tissue were both acknowledged as a phenomenon and dismissed as irrelevant because they did not fit the proper criteria and representational forms.

28. IAEA (1991a), 32.

29. Shevchouk and Gourachevsky (2001), 86–87.

30. Conclusions of the project were presented in Vienna May 21–24, 1991. Gofman (1994).

31. Shevchouk and Gourachevsky (2001), 87.

32. EC, IAEA, and WHO (1996), 9–10.

33. Ibid., 10.

34. Ibid.

35. Ibid., 17.

36. Wynne (2005), 83. In this context, the public's disagreement with scientific assessments is interpreted as "either refusal to believe, or inability to understand the quantified risk science—that is, anti-science or ignorance."

37. Chernobyl Forum (2006), 8. Similar emphasis on providing an "authoritative and definitive review of the health effects ... attributable to radiation exposure" appears in other reports, such as UNSCEAR (2008).

38. IAEA (2004).

39. IAEA (2003).

40. UNDP and UNICEF (2002). In a 2004 interview in Minsk, a local UN manager described this transformation rather cynically: "Chernobyl is a brand that one can get funding for. According to the official statistics, we do not have a problem with children's oncology, but there is a real socioeconomic problem in the affected regions. I think somebody just came up with a brilliant idea to stress the economic factors in the Chernobyl regions."

41. World Bank (2002).

42. UNDP and UNICEF (2002), 2–3, 16.

43. Ibid., 7, 37, states that morbidity in the affected territories "reflects the broader pattern of the former Soviet Union." The only radiation effect mentioned is thyroid cancer, though the report also suggests that breast cancer and some other health effects should be investigated further. With the current low doses, no health effects will be "statistically distinguishable." World Bank (2002) presents similar views, though it also includes a summary of some Belarusian studies in its attachment 3.

44. In other words, people were sick because they believed that they were sick and that the accident had made them so: "This firm belief in radiation-induced diseases ... has been a major factor in fostering a depressed health situation and an impaired state of well-being in the affected population." The situation is exacerbated by distrust in official information and "the lack of consensus among local and international specialists" on issues of radiation protection. World Bank (2002), 10.

45. Ibid., ii, 10, 46; UNDP and UNICEF (2002), 32.

46. UNDP and UNICEF (2002), 9, 16. The report notes that these efforts might not be effective for people with low incomes unless educational efforts are supplemented with measures that improve their economic status. The report suggests creating a panel of independent experts who would evaluate new research on Chernobyl issues and inform the public about the real extent of the consequences.

47. IAEA (2003). The Forum also included the Food and Agricultural Organization, the UN Environmental Program, and the UN Office for Coordination of Humanitarian Affairs.

48. Bennet, Repacholi, and Carr (2006), 69, 72, 75–76, 101–102. Some mention is made of the potentially radiation-related increase in breast cancer.

49. Ibid., 70; UNSCEAR (2000, 2008).

50. Bennet et al. (2006), 70, 93, 95, 96. This report of the "Health" group of the Chernobyl Forum explicitly notes that "the empirical studies do not support the view that the public anxiety bears a resemblance to clinical psychiatric disorders, such as phobia or psychosis," yet the experts still emphasize that "the disaster did have a psychological effect that is not limited to mental health outcomes." It is not clear from the report what expertise the "Health" group relied on in its review of the studies claiming psychological effects. Although an absence of adequate expertise would be a troubling sign, the mere presence of such experts might simply extend the basic disagreement over low-dose radiation, along with efforts at the production of public (in)visibility, into a different (expert) domain. Similar examples can be found in the history of other environmental and health risks, including, for example, the tobacco industry's turn to historians for establishing more favorable historical accounts. See Proctor and Schiebinger (2008).

51. Bennet et al. (2006), 93, 95; Chernobyl Forum (2006), 36. According to the report of the "Health" group of the Chernobyl Forum, the anxiety levels among the affected populations are higher, and the affected populations are "3–4 times more likely to report multiple unexplained physical symptoms and subjective poor health than were unaffected control groups."

52. Chernobyl Forum (2006), 41.

53. UNSCEAR (2008), 48, 65. The same statement concluded the UNSCEAR webpage about the Chernobyl accident, www.unscear.org/unscear/en/chernobyl.html. UNSCEAR (2000), 392, argues, "The tendency to attribute all problems to the accident leads to escapism, 'learned helplessness,' unwillingness to cooperate, overdependence, and a belief that the welfare system and government authorities should solve all problems."

54. These scholars note that members of the local population also had acute radiation sickness. See, e.g., Imanaka (2006).

55. Yablokov (2006); Yablokov et al. (2009).

56. Grodzinsky, Nesterenko, and Yablokov (2008); Malko, interview, Minsk, 2010.

57. Perhaps the most noticed was the U.S. publication of Yablokov et al. (2009). The reports of the European Committee on Radiation Risks have provided another visible platform for some local scientists. The committee was "formed in 1997 following a resolution made at a conference in Brussels arranged by the Green Group in the European Parliament." Busby and Yablokov (2006), 1–2, 249.

58. Indeed, Mikhail Malko, one of the leading Belarusian scholars on Chernobyl who publishes in English, explicitly connects the earlier work of John Gofman to the Belarusian scientists' efforts to demonstrate the rise in thyroid cancer. Malko, interview, Minsk, 2012. Another attempt at reexamining the established model came from the UK Committee Examining Radiation Risk from Internal Emitters (established in 2001 and disbarred in 2004), which produced two diverging reports on the issue.

59. Yablokov et al (2009). The European Committee on Radiation Risks similarly notes that that "research papers on Chernobyl effects were not being translated into English"—and were therefore ignored by WHO and UNSCEAR. Busby and Yablokov (2006), 1.

60. Busby and Yablokov (2006), 1; Yablokov (2006), 13, 17; Cohen (2008).

61. Greene (n.d). On Alice Stewart, see Greene (1999).

62. According to the critics, the UN projections of the Chernobyl health effects also appear to rely on constricted population groups and sizes. Different projects used different populations. Relative to other estimates (e.g., the estimate by Greenpeace, which was more than 20 times larger), the projections of the Chernobyl Forum were based on a particularly limited population (highly exposed liquidators, evacuees from the 30-kilometer, or 18.6-mile, zone, and inhabitants of the most heavily contaminated areas). Imanaka (2008), 8, writes, "This difference reflects the fact that the number of cancer deaths largely depends on the risk model and the size of population used by the evaluator." His own estimate is for "a total of 20,000–60,000 cancer deaths."

63. Mikhail Malko pointed out in an interview that we can observe the consequences of Chernobyl contamination because the collective dose after Chernobyl was higher than it was in Hiroshima and Nagasaki, and the number of radiation-induced health effects is calculated as a function of the collective dose of irradiation. Explaining that the duration of the latent period can be lessened if the number of exposed people increases, Malko referred to the work of John Gofman. In his published work, Malko (1998b), 10, maintains a similar connection between what Belarusian scientists have shown in the case of children's thyroid cancer and Gofman's work: "The [Belarusian] specialists have managed to prove the validity of this idea by Prof. J. Gofman in the case of thyroid cancer, thus making a significant contribution to the study of radiation effects on the organism."

64. Gofman (1990, 1994). For Gofman's disagreement with the U.S. Atomic Energy Commission over radiation safety guidelines, see Boudia (2007).

65. Gofman (1990), Foreword-1. Gofman (1994) offers a number of organizational suggestions to minimize researchers' bias.

66. Nesterenko (1998) argues that these measurements were later used to justify lifting radiation protection measures in areas with a level of surface contamination of 1–5 curies per square kilometers. These dose estimates were lower than the results of measurements by Belrad. Vassily Nesterenko, interview, Minsk, 2004.

67. Belbéoch (1998).

68. Yablokov et al. (2009), 36. Yablokov explains that the lack of specific scientific protocols usually means that "there was no statistical processing of the received data."

69. Ibid., 320.

70. Grodzinsky et al. (2008), 29, 30. In some cases, the researchers note that these effects, including noncancer diseases, are then confirmed by the Japanese studies. For example, Malko called my attention to the following overview of cancer and noncancer diseases among the atomic bomb survivors, which notes an increased risk of circulatory, respiratory and digestive diseases; Ozasa, Shimizu, Suyama, Kasagi, Soda, Grant, Sakata, Sugiyama, and Kodama (2012).

71. Shevchouk and Gourachevsky (2003), 93. This national report specifically notes that the Belarusian government was the first among the governments of the three most affected countries to turn to this approach.

72. World Bank (2002), 10.

73. UNDP and UNICEF (2002), 30.

74. According to UNDP and UNICEF (2002), 61, "the main downside" is that the international health recuperation programs that arrange for the Belarusian children from the affected territories to spend time abroad "may perpetuate inaccurate and negative stereotypes about life in the affected areas, both in the minds of the young people and in the host community."

75. World Bank (2002), 53.

76. Establishment of the CORE program followed the UNDP and UNICEF report and the World Bank report. It also followed reports by the heads of missions of the ambassadors from the European Union about their visits to the affected territories (April 2001 and May 2003) and the report of the experience of the Ethos project, which was discussed during the international seminar held in Stolin, Belarus (November 15–16, 2001). The Ethos project, sponsored by the European Union, worked with various groups of the local population to address an aspect of their lives related to radiation protection. CORE (2003).

77. ICRIN (2004b), 5-7; ICRIN (2004a), 39. The concept of ICRIN was elaborated by the Swiss Agency for Development and Cooperation; the work of the network was endorsed by the UN Inter-Agency Task Force on Chernobyl in 2003 and coordinated with the Chernobyl Forum.

78. CORE (2003), 2.

79. ICRIN (2004a), 32; Belookaya (2004a), 3–5.

80. CORE's offices were right next door to the offices of the State Committee on Chernobyl.

81. Zoya Trofimchik, interview, Minsk, 2004.

82. The program included agronomists who gave advice on growing uncontaminated produce, and at least one of the senior managing staff members held an advanced degree in science. Beyond that, there was no involvement of the local scientific establishment or any science perspective other than that of the international nuclear experts.

83. Topçu (2013).

84. Belrad personnel, interviews, Minsk, 2010, 2012.

85. Evgeni Konoplya, interview, Minsk, 2005. All subsequent quotes are from this interview.

86. Boudia (2013), 82.

87. Gofman (1994).

88. Belookaya, Korytko, Melnov, Tegako, and Chernenok (2002), 6; see also Yablokov et al. (2009), 325.

89. Tamara Belookaya, interview, Minsk, 2005.

90. Alexey Nesterenko, interview, Minsk, 2010.

91. UNDP and UNICEF (2002), 61.

92. Williams, Becker, Demidchik, Nagataki, Pinchera, and Tronko (1996), 212.

93. Hecht (2012), 43.

94. Wynne (1991), 120. Imanaka (2008), 18.

95. Imanaka (2008), 18.

Chapter 6

1. *SB*, February 26, 1999.

2. Bandazhevsky (1997). Bandazhevsky's work is cited extensively in Yablokov et al. (2009).

3. Bandazhevsky (1999, 2003).

4. Nesterenko and Nesterenko (2006), 196; Nesterenko and Nesterenko (2009), 304.

5. Bandazhevsky's critique appeared in his report on the directions of research in the Institute of Radiation Medicine of the Ministry of Health. His concerns about the state's policies of rehabilitation and agricultural production in the most affected areas appear again in the conclusion of Bandazhevsky (2003), 490, in which he argues that studying the effects of internal accumulation of Cesium-137 on pathogenesis in children is

an urgent need, as radiocontaminated agricultural land is being increasingly cultivated and radiocontaminated food is circulating countrywide [in Belarus]. Schoolchildren in contaminated areas received radiologically clean food free of charge in school canteens and spent a month in a sanatorium, in a clean environment, each year. For reasons of economy the annual sanatorium stay has been shortened, and communities in some contaminated areas have been classified as "clean," thus ending the supply of clean food from the state.

6. Amnesty International (2005).

7. *Nasha Niva*, April 23, 2007.

8. Konoplya and Rolevich (1996a), 76; Matsko and Imanaka (1998), 28–39; Shevchouk and Gourachevsky (2001). The Institute of Radiobiology and the Institute of Radioecological Problems were institutes of the National Academy of Sciences of Belarus. The Institute of Agricultural Radiology later became the Institute of Radiology under the State Committee on Chernobyl. Other selected institutes with departments or laboratories conducting Chernobyl-related research included the Institute of Nuclear Power Energy; the Belarusian Scientific Research Institute of Hematology and Blood Transfusion; the Institute of Oncology and Medical Radiology; the Institute of Soil Science and Agrochemistry; the Belarus Center of Medical Technologies, Information, Direction and Economy of Public Health; and the Institute for Genetics and Cytology.

9. Belarus inherited the Soviet vertical concentration of power, and the state administration has been particularly centralized since the election of the first (and as I am writing this, only) Belarusian president in 1994. Consequently, this analysis refers to the state as a single and rather coherent actor and not as a set of different agencies with potentially competing interests, which would be more appropriate in other contexts.

10. Nesvetailov (1995), 858.

11. Josephson (1999).

12. Tamara Belookaya, interview, Minsk, 2005.

13. Nesvetailov (1995), 858.

14. Evgeni Konoplya, interview, Minsk, 2005; see also NASB (1999). Konoplya was the director of the Institute of Radiobiology until 2009.

15. Marples (1996a), 104–109; NASB (2010). Demidchik identified the clinical-biological specificity of radiation-induced thyroid cancer in children.

16. Belarus Ministry of Emergencies (2011), 21.

17. Konoplya and Rolevich (1996a), 77. In addition to this plan for scientific research and development, in 1990 the republic adopted a separate program that outlined the monitoring and forecasting of the radiological situation.

18. Ibid., 76.

19. Ibid., 77; Matsko and Imanaka (1998), 34.

20. Matsko and Imanaka (1998); Shevchouk and Gourachevsky (2001).

21. Nesvetailov (1995), 860, 863. The rate of inflation in 1994 was about 2,100 percent.

22. Konoplya and Rolevich (1996a), 77.

23. Nesvetailov (1995), 864.

24. For example, a presidential decree, "On Improving State Management in the Area of Science," issued on March 5, 2002, adjusted the hierarchy and responsibilities of some science-related administrative bodies, including the National Academy of Sciences. Korshunov, Artuhin, Elsukov, Kostukovich, Nikonovich, Rudenkov, Tamashevich, and Hartonik (2006).

25. Konoplya and Rolevich (1996a), 77.

26. Nesvetailov (1995), 854–856.

27. Yet Nesvetailov (1995), 871n49 wrote that mitigating Chernobyl's consequences was "a major priority over which the country has had no choice."

28. Nesvetailov (1995), 866–867, also observed "strong evidence of a growing western orientation in Belarusian science at the level of the state, Academy [of Sciences], and institutes."

29. Rajan (2002). Fortun (2001) offers the concept of *continuing disaster*.

30. Konoplya (1996), 4.

31. *Sovetskaya Byelorussiya*, February 9, 1989.

32. Shevchouk and Gourachevsky (2001), 79.

33. *Narodnaya Gazeta*, January 24, 1997. The clinic in Aksakovshchina provided health-care services for the Chernobyl-affected populations. It was first designed to accommodate 190 patients (95 children and 95 adults) but was then expanded to 250, half of which were endocrine patients.

34. Ibid.

35. Former employee of the Institute of Radiation Medicine and Endocrinology in Minsk, interview, Minsk, 2005. On July 5, 1994, the clinic in Aksakovshchina was

transformed into the Institute of Endocrinology, then transformed back after 43 days. *Narodnaya Gazeta*, January 24, 1997.

36. The University was first established in 1992 as the Sakharov International College of Radioecology, a subdivision of Belarusian State University. In 1994, it became an independent institute with two departments. Shevchouk and Gourachevsky (2001), 97.

37. Former employee of the Institute of Radiation Medicine and Endocrinology in Minsk, interview, Minsk, 2005. The establishment was originally part of the Institute for Radiation Medicine but later became independent. The original building was also close to the hospital for the president's administration.

38. Belarus Ministry of Emergencies (2011), 23–25; Shevchouk and Gourachevsky (2006), 89.

39. Shevchouk and Gourachevsky (2006), 89. These directions for research were also specified in the 2006–2010 Chernobyl Program; see Belarus Ministry of Emergencies (2011), 24.

40. *Sovetskaya Byelorussiya*, February 26, 1999.

41. Evgeni Konoplya, interview, Minsk, 2005.

42. The creation of the center was specified by the 1990–1995 Chernobyl Program. Yet when the center opened in 1990, it did so without its own building. The construction of the building for the center stopped altogether in 1994 because of insufficient funds. Efforts to renew the construction began again in the late 1990s, under the patronage of the president. *Sovetskaya Byelorussiya*, February 20, 1997, and June 11, 1998.

43. *BelaPAN*, July 5, 2007, http://naviny.by/rubrics/society/2007/07/05/ic_news_116_273394.

44. Nesterenko (n.d.).

45. Belarus Ministry of Emergencies, (2011), 39.

46. Shevchouk and Gourachevsky (2006), 69.

47. This point about the role of annual examinations was emphasized to me during a 2004 interview in the Department of Science, Ministry of Health. For their examination, residents can go to their local medical facilities or travel to the administrative center of their region. A number of groups of individuals are subject to annual medical examinations: cleanup workers, evacuees, residents of the affected areas, children of the most heavily exposed individuals (cleanup workers, evacuees, and residents of the zones of primary and secondary resettlement), and children under 18 with radiation-induced diseases that had been established as such (typically

leukemia and thyroid cancer but also other cancers). Shevchouk and Gourachevsky (2006), 69–70; Belarus Ministry of Emergencies, (2011), 38–39.

48. Marples (1996a).

49. Former physician of the outpatient center of the Institute of Radiation Medicine in Minsk, interview, Minsk, 2004. The next quote is from the same interview. The outpatient center no longer performed any Chernobyl-related functions at the time of this interview, though some of its personnel still remained. For a discussion of the inadequacies of the general healthcare system, especially in the affected rural areas, see Marples (1996a), UNDP and UNICEF (2002), and World Bank (2002).

50. Marples (1996a); Tamara Belookaya, interview, Minsk, 2005. Although the registry is not a perfect reflection of the reality, a blanket invalidation of the Belarusian data because of this would make even more data invisible.

51. Tamara Belookaya, interview, Minsk, 2005.

52. Shevchouk and Gourachevsky (2006), 71; Belarus Ministry of Emergencies, (2011), 40. Although these national reports note that this group is at risk, there is no separate category for them, and it is not clear whether the screening for them remains focused on thyroid pathology.

53. Bowker and Star (1999). Ensuring that categories are consistent and uniform is work that must be done in the actual practice of using categories. See chapter 4 for a discussion on aligning categories.

54. NRC (2006), 202. A separate registry of cancers has been functioning in the republic since 1970s.

55. Tamara Belookaya, interview, Minsk, 2005.

56. For more on this, see Brown (2013).

57. State power relies not only on the collection of data but also on strategic ignorance and a lack of usable data in particular areas. These dimensions of power, as Mathews (2005) demonstrates, tended to be overlooked by social sciences influenced by Michel Foucault's concept of power/knowledge.

58. Vassily Nesterenko, interview, Minsk, 2004; Nikitchenko (1999).

59. Galina Bandazhevskaya, interview, Minsk, 2004. All subsequent quotes are from this interview.

60. Evgeni Konoplya, interview, Minsk, 2005.

61. Bandazhevskaya's view of the radiation factor is different from that advocated by the IAEA experts who make no distinction between internal and external exposures.

62. Reports from different years are fairly consistent in their descriptiosn of these trends, even though some researchers (and members of Chernobyl-related organizations) that I have interviewed distinguished among reports from different years, noting that some are far more descriptive than others.

63. Konoplya and Rolevich (1996a), 53; Shevchouk and Gourachevsky (2001), 77; Shevchouk and Gourachevsky (2003), 51–53; Shevchouk and Gourachevsky (2006), 43.

64. Konoplya and Rolevich (1996a), 53.

65. Shevchouk and Gourachevsky (2006), 43. The report refers to "a linear dependence between accumulated radiation dose and realized relative risk of breast cancer development" for the female population of the Gomel region.

66. Konoplya and Rolevich (1996a), 85.

67. Shevchouk and Gourachevsky (2001), 106.

68. Ibid., 107. The assessment in question is UNSCEAR (2000). The Belarusian report points out that although cancers have traditionally been considered the main health effect of radiation, more data have appeared in the last few years indicating the "radiation origin" of a range of noncancer diseases.

69. Shevchouk and Gourachevsky (2003), 52.

70. Shevchouk and Gourachevsky (2001), 23, 106–107.

71. Shevchouk and Gourachevsky (2003), 48, 51, 52; Shevchouk and Gourachevsky (2006), 43.

72. Shevchouk and Gourachevsky (2006), 92,101, 102. The troubling state of children's health in the affected areas is addressed outside the discussion of causality. This report, like the earlier reports, calls for long-term radiation-epidemiological studies to demonstrate the role of radiation in the increase of the number of cancers and noncancer diseases. In conclusion, the report argues that "efforts of the Belarusian scientists should be directed to obtaining the reliable data recognized by the international community and capable to be a basis for ... effective actions at both national and international levels."

73. *Gomel'skaya Pravda*, March 13, 2009. See also *Belta*, December 27, 2012, http://news.tut.by/health/328251.html. Kenigsberg is one of the coauthors, with Evgeni Petryaev and others, of the 1995 revised concept of radiation protection.

74. Former physician of the outpatient center of the Institute of Radiation Medicine in Minsk, interview, Minsk, 2004.

75. *Vecherni Grodno*, April 26, 2009.

76. Crenson (1971).

77. Physician of the Center for Cancers of the Thyroid Gland, interview, Minsk, 2004. All subsequent quotes are from this interview.

78. I have encountered similar reactions among experts from the Sakharov International Ecological University, the former Center for Radiation Medicine, the State Pediatric Center for Oncology, and even the science department of the Ministry of Health.

79. One journal devoted to Chernobyl research, *Ecological Anthropology*, was being published on the basis of the proceedings of an annual Chernobyl conference organized by the Belarusian Committee "Children of Chernobyl" (it was published under the name *Chernobyl Catastrophe* from 1992 to 1995). According to its editor, Tamara Belookaya, the fact that scientific publications from the neighboring Ukraine and Russia rarely made it to researchers in Belarus explained its popularity.

80. Interview, Minsk, August 2011.

Conclusion

1. Norman (1998).

2. Hess (2007); Langston (2011); McGarity and Wagner (2010); Michaels (2008); Murphy (2006); Oreskes and Conway (2010); Proctor (1995); Proctor and Schiebinger (2008).

3. Edwards (2010); Fortun (2001); Oreskes and Conway (2010).

4. For example, in April 2010 the eruption of a volcano in Iceland and the consequent release of volcanic ash led to flights being halted in Europe. The airlines were quick to describe this situation as "not sustainable" and demanded acceptance of higher degrees of risk. *New York Times*, April 10, 2011.

5. Within days after the Fukushima accident, the annual allowable exposure level for nuclear workers was raised from 100 to 250 millisieverts. In April, the exposure limit for the general population, including children, was raised from 1 millisievert per year to 20, perhaps in an attempt to avoid more evacuations and cleanup measures. After much public outcry, the original limit for children was restored in May, yet it applied only to children inside school buildings. *New York Times*, August 8, 2011.

6. *PBS Newshour*, November 10, 2011.

7. *Marketplace*, NPR, March 11, 2013.

8. Brown (2007).

9. DeLuca (1999), 1-3. For example, one of the most powerful images from the 2010 documentary film *Gasland* on the effects of hydraulic fracturing ("fracking") is the

image of tap water bursting into flames. Such a dramatic demonstration might be difficult to counteract with lengthy explanations or counterarguments. See also Peeples (2011).

10. Helen Evans, "Nuage Vert," May 2008, http://hehe.org.free.fr/hehe/texte/nv/ index.html. I owe this example to Geoffrey Bowker.

11. Langston (2011), 166.

12. Mirny (2000, 2009).

Appendix

1. Charmaz (2006); Clarke (2005); Glaser and Strauss (1987); Strauss (1987); Strauss and Corbin (1990).

2. Glaser and Strauss (1987), 61.

3. *Narodnaya Volya*'s significance as an oppositional newspaper is illustrated by the fact that 250,000 of its copies were seized on March 3, 2006, before the presidential election—just as the copies were being delivered from the publisher in Smolensk, Russia. On the following Monday, the printer suddenly canceled the paper's contract, according to the editor, Iosif Seredich. *International Herald Tribune*, Sunday, March 19, 2006.

4. In 1990, *Sovetskaya Byelorussiya* published full stenographic discussions of the Chernobyl question at the sessions of the Supreme Soviet of the Byelorussian Soviet Socialist Republic (running up to eight full newspaper pages).

References

Alexievich, Svetlana. 1999. *Voices from Chernobyl: Chronicle of the Future*. London: Aurum.

Almklov, Petter G. 2008. "Standardized Data and Singular Situations." *Social Studies of Science* 38 (6): 873–897.

Amnesty International. 2005. "Prisoner of Conscience Professor Yury Bandazhevsky Is Free!" August 5. http://www.amnestyusa.org/our-work/latest-victories/prisoner-of-conscience-professor-yury-bandazhevsky-is-free.

ASPA Russia-Belarus. 2009. *Atlas sovremennyh i prognoznyh aspektov posledstvii avarii na Chernobyl'skoi AES na postradavshih territoriyah Rossii i Belarusi* [Atlas of Contemporary and Forecast Aspects of the Consequences of the Accident at the Chernobyl NPP for the Affected Territories of Russia and Belarus]. Moscow: Ministry of Emergencies of Russia.

Babenko, Vladimir I. 2003. *Kak zashchitit'sebya i svoego rebenka ot radiacii: Posobie dlya roditeley* [How to Protect Yourself and Your Child from Radiation: A Guide for Parents]. Minsk, Belarus: Institute of Radiation Safety "Belrad" (hereafter, "Belrad").

Bakhtin, Mikhail. 1984. *Problems of Dostoevsky's Poetics*. Minneapolis: University of Minnesota Press.

Bakhtin, Mikhail. 1981. *The Dialogic Imagination: Four Essays*. Austin: University of Texas Press.

Bandazhevskaya, Galina S., Vassily B. Nesterenko, Vladimir I. Babenko, T. V. Yerkovich, and Yuri I. Bandazhevsky. 2004. "Relationship between Caesium (137Cs) Load, Cardiovascular Symptoms, and Source of Food in 'Chernobyl' Children: Preliminary Observations after Intake of Oral Apple Pectin." *Swiss Medical Weekly* 134 (January 1): 725–729.

Bandazhevsky, Yuri. 1997. *Pathology and Physiology of the Incorporated Ionizing Radiation*. Gomel, Belarus: Gomel Medical Institute.

Bandazhevsky, Yuri. 1999. *Pathology of Incorporated Ionizing Radiation*. Minsk: Belarus Technical University.

Bandazhevsky, Yuri. 2000. *Biomedical Effects of Radiocesium Incorporated into the Organism*. Minsk, Belarus: Belrad.

Bandazhevsky, Yuri. 2001. *Radioactive Caesium and Intrauterine Fetus Development*. Minsk, Belarus: Belrad.

Bandazhevsky, Yuri. 2003. "Chronic Cs-137 Incorporation in Children's Organs." *Swiss Medical Weekly* 133 (January 1): 488–490.

Beck, Ulrich. 1992. *Risk Society: Towards a New Modernity*. Newbury Park, CA: Sage.

Beck, Ulrich. 1995a. *Ecological Enlightenment: Essays on the Politics of the Risk Society*. Amherst, NY: Humanity Books.

Beck, Ulrich. 1995b. *Ecological Politics in an Age of Risk*. Cambridge, UK: Polity.

Beck, Ulrich. 1999. *World Risk Society*. Cambridge, UK: Polity.

Beck, Ulrich, Anthony Giddens, and Scott Lash, eds. 1994. *Reflexive Modernization: Politics, Tradition and Aesthetics in the Modern Social Order*. Stanford, CA: Stanford University Press.

Belarus Committee on the Problems of the Consequences of the Catastrophe at the Chernobyl Nuclear Power Plant (Belarus Committee on Chernobyl) and United Nations Development Programme (UNDP). 2005. *2005 Calendar.*

Belarus Ministry of Emergencies. 2011. *Chetvert'veka posle Chernobyl'skoi katastrofy: Itogi i perspektivy preodoleniya* [A Quarter of a Century after the Chernobyl Catastrophe: Results and Prospects of Mitigation]. National Report. Minsk, Belarus: Institute of Radiology.

Belarus Ministry of Health. 2002. *Zhizn' posle Chernobylya: 16 let spustya* [Life after Chernobyl: 16 Years Later]. Proceedings of the Minsk Oncology Clinic Conference, June 20.

Belbéoch, Bella. 1998. "Western Responsibility regarding the Health Consequences of the Chernobyl Catastrophe in Belarus, the Ukraine and Russia." http://www.dissident-media.org/infonucleaire/western_responsibility.html.

Belookaya, Tamara V. 2004a. *Guide on the Organization of the Process of the Scientific Research and Information Network on Chernobyl: Interactive Study of Information Needs of the Populations Affected by the Chernobyl Accident*. Minsk, Belarus: Belarusian Committee "Children of Chernobyl."

Belookaya, Tamara V. 2004b. *Information Needs Assessment of the Populations Affected by the Chernobyl Catastrophe*. Report, March 15–April 26.

Belookaya, Tamara V. 2004c. "An Interactive Study of the Non-Governmental Organization Belarusian Committee 'Children of Chernobyl'." In *An Information Needs Assessment of the Chernobyl-Affected Population in the Republic of Belarus*. Minsk, Belarus: Unipack.

Belookaya, Tamara V. 2004d. "On Information Strategies for the Chernobyl Regions: Results of the Interactive Study of Information Needs of the Affected Populations." Unpublished paper.

Belookaya, Tamara V., S. S. Korytko, S. B. Melnov, L. I. Tegako, and D. A. Chernenok. 2002. "The Problems of Effects of Small Doses of Ionizing Radiation." *Ekologicheskaya Antropologiya*, annual edition, 5-45. Minsk: Belarussian Committee "Children of Chernobyl"

Bennet, B., M. Repacholi, and Z. Carr, eds. 2006. *Health Effects of the Chernobyl Accident and Special Health Care Programmes*. Report of the UN Chernobyl Forum Expert Group "Health." Geneva: World Health Organization.

Bertell, Rosalie. 1999. "Avoidable Tragedy Post-Chernobyl: A Critical Analysis." *Journal of Humanitarian Medicine* 2 (3): 21–28.

Bertell, Rosalie. 2005. "Comments on the Press Release 'Chernobyl: The True Scale of the Accident'." September 10. http://www.iicph.org.

Billig, Michael. 1998. *Talking of the Royal Family*. 2nd ed. London: Routledge.

Boudia, Soraya. 2007. "Global Regulation: Controlling and Accepting Radioactivity Risks." *History and Technology* 23 (4): 389–406.

Boudia, Soraya. 2013. "From Threshold to Risk: Exposure to Low Doses of Radiation and Its Effects on Toxicants Regulation." In *Toxicants, Health, and Regulation since 1945*, edited by Soraya Boudia and Nathalie Jas, 71–87. London: Pickering & Chatto.

Boudia, Soraya, and Nathalie Jas, eds. 2013. *Toxicants, Health, and Regulation since 1945*. London: Pickering & Chatto.

Bowers, J. 1992. "The Politics of Formalism." In *Contexts of Computer-Mediated Communication*, edited by Martin Lea, 232–261. Hassocks, UK: Harvester.

Bowker, Geoffrey C. 2005a. "Cognitive Work and Material Practice." *Journal of the Learning Sciences* 14 (1): 157–160.

Bowker, Geoffrey C. 2005b. *Memory Practices in the Sciences*. Cambridge, MA: MIT Press.

Bowker, Geoffrey C., and Susan Leigh Star. 1999. *Sorting Things Out: Classification and Its Consequences*. Cambridge, MA: MIT Press.

Briggs, Charles L. 2003. "Why Nation States and Journalists Can't Teach People to Be Healthy: Power and Pragmatic Miscalculation in Public Discourses on Health." *Medical Anthropology Quarterly* 17 (3): 287–321.

Briggs, Charles L., with Clara Mantini-Briggs. 2003. *Stories in the Time of Cholera: Racial Profiling during a Medical Nightmare*. Berkeley: University of California Press.

Brown, Kate. 2013. *Plutopia: Nuclear Families, Atomic Cities, and the Great Soviet and American Plutonium Disasters*. New York: Oxford University Press.

Brown, Phil. 1992. "Popular Epidemiology and Toxic Waste Contamination: Lay and Professional Ways of Knowing." *Journal of Health and Social Behavior* 33: 267–281.

Brown, Phil. 1997. "Popular Epidemiology Revisited." *Current Sociology* 45(3): 137–156.

Brown, Phil. 2007. *Toxic Exposures: Contested Illnesses and the Environmental Health Movement*. New York: Columbia University Press.

Brown, Phil, Steve Kroll-Smith, and Valerie Gunter. 2000. "Knowledge, Citizens, and Organizations: An Overview of Environments, Diseases, and Social Conflict." In *Illness and the Environment: A Reader in Contested Medicine*, edited by S. Kroll-Smith, P. Brown, and V. J. Gunter, 9–25. New York: New York University Press.

Busby, Chris C., and Alexey V. Yablokov, eds. 2006. *Chernobyl: 20 Years On. Health Effects of the Chernobyl Accident*. Report of the European Committee on Radiation Risk (ECRR). 2nd edition. Aberystwyth, UK: Green Audit Press.

Busch, Lawrence. 2011. *Standards: Recipes for Reality*. Cambridge, MA: MIT Press.

Carson, Rachel. 1962. *Silent Spring*. Greenwich, CT: Fawcett Crest.

Charmaz, Kathy. 2006. *Constructing Grounded Theory: A Practical Guide Through Qualitative Analysis*. London: Sage.

Chernobyl Forum. 2006. *Chernobyl's Legacy: Health, Environmental and Socioeconomic Impacts, and Recommendations to the Governments of Belarus, the Russian Federation, and Ukraine*. 2nd rev. ed. http://www.iaea.org/Publications/Booklets/Chernobyl/chernobyl.pdf.

Clarke, Adele. 2005. *Situational Analysis: Grounded Theory after the Postmodern Turn*. Thousand Oaks, CA: Sage.

Cohen, Bernard. 2008. "The Linear No-Threshold Theory of Radiation Carcinogenesis Should Be Rejected." *Journal of American Physicians and Surgeons* 13 (3): 70–76.

Cooperation for Rehabilitation (CORE). 2003. "Declaration of Principles of the CORE Programme 'CO-operation for RE-habilitation of living conditions in Chernobyl Affected Areas in Belarus.'"

Corbin, Juliet M., and Anselm L. Strauss. 1988. *Unending Work and Care: Managing Chronic Illness at Home*. San Francisco, CA: Jossey-Bass.

Cox, Robert. 2010. *Environmental Communication and the Public Sphere*. 2nd edition. Los Angeles, CA: Sage.

Crenson, Matthew A. 1979. *The Un-Politics of Air Pollution: A Study of Non-Decision-making in the Cities*. Baltimore, MD: Johns Hopkins University Press.

DeLeo, Maryann, dir. and prod. 2003. *Chernobyl Heart* [film].

DeLuca, Kevin M. 1999. *Image Politics: The New Rhetoric of Environmental Activism*. New York: Guilford Press.

Dunwoody, Sharon, and Hans Peter Peters. 1992. "Mass Media Coverage of Technological and Environmental Risks: A Survey of Research in the U.S. and Germany." *Public Understanding of Science* 1(2): 199–230.

Edelstein, Michael. 1988. *Contaminated Communities: The Social and Psychological Impacts of Residential Toxic Exposure*. Boulder, CO: Westview Press.

Eder, Klaus. 1996. *The Social Construction of Nature*. London: Sage.

Edwards, Paul N. 2010. *A Vast Machine: Computer Models, Climate Data, and the Politics of Global Warming*. Cambridge, MA: MIT Press.

Epstein, Steven. 1996. *Impure Science: AIDS, Activism, and the Politics of Knowledge*. Berkeley: University of California Press.

European Commission (EC), International Atomic Energy Agency (IAEA), and World Health Organization (WHO). 1996. *One Decade after Chernobyl: Summing Up the Consequences of the Accident*. Proceedings of an International Conference, Vienna, April 8–12.

Fischer, David. 1997. *History of the International Atomic Energy Agency: The First Forty Years*. Vienna: IAEA.

Fortun, Kim. 2001. *Advocacy after Bhopal: Environmentalism, Disaster, New Global Orders*. Chicago: University of Chicago Press.

Fortun, Kim. 2004. "From Bhopal to the Informating of Environmentalism: Risk Communication in Historical Perspective." *Osiris* 19: 283–296.

Fowlkes, Martha R., and Patricia Y. Miller. 1987. "Chemicals and Community at Love Canal." In *The Social and Cultural Construction of Risk: Essays on Risk Selection and Perception*, edited by B. B. Johnson and V. T. Covello, 55–78. Boston: Reidel.

Fujimura, Joan. H. 1987. "Constructing 'Do-Able' Problems in Cancer Research: Articulating Alignment." *Social Studies of Science* 17 (2): 257–293.

Gamson, William A., and Andre. Modigliani. 1989. "Media Discourse and Public Opinion on Nuclear Power: A Constructionist Approach." *American Journal of Sociology* 95 (1): 1–37.

Gitlin, Todd. 1980. *The Whole World Is Watching: Mass Media in the Making and Unmaking of the New Left*. Berkeley: University of California Press.

Glaser, Barney G., and Anselm L. Strauss. 1987. *The Discovery of Grounded Theory: Strategies for Qualitative Research*. New York: Aldine de Gruyter.

Goffman, Erving. 1974. *Frame Analysis: An Essay on the Organization of Experience*. New York: Harper Colophon.

Gofman, John W. 1990. *Radiation-Induced Cancer from Low-Dose Exposure: An Independent Analysis*. San Francisco: Committee for Nuclear Responsibility.

Gofman, John W. 1994. *Chernobyl'skaya avariya: Tadiacionnye posledstviya dlya nastoyashchego i buduschego pokolenii* [Chernobyl Accident: Radiation Consequences for This and Future Generations]. Minsk, Belarus: Vysheishaya Shkola.

Goodwin, Charles. 1994. "Professional Vision." *American Anthropologist* 96 (3): 606–633.

Gould, Kenneth A. 1993. "Pollution and Perception: Social Visibility and Local Environmental Mobilization." *Qualitative Sociology* 16 (2): 157–178.

Greene, Gayle. 1999. *The Woman Who Knew Too Much: Alice Stewart and the Secrets of Radiation*. Ann Arbor: University of Michigan Press.

Greene, Gayle. n.d. "Science with a Skew: The Nuclear Power Industry after Chernobyl and Fukushima." *Asia-Pacific Journal*, http://www.japanfocus.org/-Gayle-Greene/3672.

Grodzinksy, Dimintro M., Vassily B. Nesterenko, and Alexey V. Yablokov. 2008. "Ne katastrofa, ne avariya, a prosto pozhar? Zamechaniya na polyah doklada OON 2002g." [Not a Catastrophe, Not an Accident, Just Fire? Comments on the Margins of the 2002 UN Report]. In Vassily Nesterenko. *Sbornik statei i dokladov, 2001–2008* [Collection of Articles and Reports], 25–33. Minsk: Belrad.

Haraway, Donna. 1991. "Situated Knowledges: The Science Question in Feminism and the Privilege of Partial Perspective." In *Simians, Cyborgs, and Women: The Reinvention of Nature*, 183–201. New York: Routledge.

Hecht, Gabrielle. 1998. *The Radiance of France: Nuclear Power and National Identity after World War II*. Cambridge, MA: MIT Press.

Hecht, Gabrielle. 2006. "Negotiating Global Nuclearities: Apartheid, Decolonization, and the Cold War in the Making of the IAEA." *Osiris* 21 (July): 25–48.

Hecht, Gabrielle. 2012. *Being Nuclear: Africans and the Global Uranium Trade.* Cambridge, MA: MIT Press.

Hess, David J. 2007. *Alternative Pathways in Science and Industry: Activism, Innovation, and the Environment in an Era of Globalization.* Cambridge, MA: MIT Press.

Hoffman, Susanna M., and Anthony Oliver-Smith, eds. *Catastrophe & Culture: The Anthropology of Disaster.* Santa Fe, NM: School of American Research Press.

Imanaka, Tetsuji. 2008. "What Happened at That Time?" In *Multi-Side Approach to the Realities of the Chernobyl NPP Accident: Summing Up of the Consequences of the Accident Twenty Years After,* Report, 1–19. Kyoto, Japan: Kyoto University Research Reactor Institute. http://www.rri.kyoto-u.ac.jp/NSRG/reports/kr139/pdf/kr139.pdf.

International Atomic Energy Agency (IAEA). 1991a. *The International Chernobyl Project: Assessment of Radiological Consequences and Evaluation of Protective Measures.* An Overview. Vienna.

International Atomic Energy Agency (IAEA). 1991b. *The International Chernobyl Project: Assessment of Radiological Consequences and Evaluation of Protective Measures.* Summary brochure. Vienna.

International Atomic Energy Agency (IAEA). 1991c. *The International Chernobyl Project: Assessment of Radiological Consequences and Evaluation of Protective Measures.* Technical report. Vienna.

International Atomic Energy Agency (IAEA). 2003. "Forum Sharpens Focus on Human Consequences of Chernobyl Accident." February 6. http://www.iaea.org/newscenter/features/chernobyl-15/forum_launched.shtml.

International Atomic Energy Agency (IAEA). (2004). "Chernobyl: Clarifying Consequences." http://www.iaea.org/NewsCenter/News/2004/consequences.html.

International Chernobyl Research and Information Network (ICRIN). 2004a. *An Information Needs Assessment on the Chernobyl-Affected Population in the Republic of Belarus.* Minsk, Belarus: Unipack.

International Chernobyl Research and Information Network (ICRIN). 2004b. *International Chernobyl Research and Information Network.* Minsk, Belarus: Unipack.

Irwin, Alan. 1995. *Citizen Science: A Study of People, Expertise and Sustainable Development.* New York: Routledge.

Jasanoff, Sheila. 1990. *The Fifth Branch: Science Advisers as Policymakers.* Cambridge, MA: Harvard University Press.

Jasanoff, Sheila. 2003. "Technologies of Humility: Citizen Participation in Governing Science." *Minerva* 41: 223–244.

Josephson, Paul R. 1999. *Red Atom: Russia's Nuclear Power Program from Stalin to Today*. New York: W. H. Freeman.

Kenik, Ivan A., ed. 1998. *Belarus and Chernobyl: The Second Decade*. Minsk, Belarus: Ministry of Emergencies.

Konoplya, Evgeni F. 1996. *"Nauka i ee rol'v reshenii Chernobyl'skih problem"* [Science and Its Role in Solving Chernobyl Problems]. In *Desyat'let posle Chernobyl'skoi katastrofu: Nauchnye aspekty problemy* [Ten Years after the Chernobyl Catastrophe: Scientific Aspects of the Catastrophe], 3–5. Abstracts of the Scientific Conference, Academy of Sciences of Belarus, February 28–29.

Konoplya, Evgeni F., and I. V. Rolevich, eds. 1996a. *The Chernobyl Catastrophe Consequences in the Republic of Belarus*. National Report. Minsk: Academy of Sciences of Belarus.

Konoplya, Evgeni F., and I. V. Rolevich, eds. 1996b. *Ecologicheskie, mediko-biologicheskie i social'no-ekonomicheskie posledstviya katastrofy na ChAES v Belarusi* [Ecological, Medical-Biological, and Socioeconomic Consequences in Belarus of the Catastrophe at the Chernobyl NPP]. Minsk: Institute of Radiobiology of the Academy of Sciences of Belarus.

Korshunov, A.N., M.I. Artuhin, V.P. Elsukov, N.N. Kostukovich, S.V. Nikonovich, V.M. Rudenkov, V.N. Tamashevich, I.A. Hartonik. 2006. *On the State and Perspectives of the Development of Science in the Republic of Belarus, Based on the Results from 2005 and the Period from 2001-05*. Analytical Report. Minsk: Belarusian Institute of System Analysis.

Kroll-Smith, Steve, Phil Brown, and Valerie J. Gunter, eds. 2000. *Illness and the Environment: A Reader in Contested Medicine*. New York: New York University Press.

Kroll-Smith, Steve, and H. Hugh Floyd. 1997. *Bodies in Protest: Environmental Illness and the Struggle over Medical Knowledge*. New York: New York University Press.

Kuchinsky, Gennadi. 1988. *Psihologiya vnutrennego dialoga* [Psychology of Internal Dialogue]. Minsk, Belarus: University Press.

Lampland, Martha. 2010. "False Numbers As Formalizing Practices." *Social Studies of Science* 40 (3): 377–404.

Lampland, Martha, and Susan Leigh Star. 2009. *Standards and Their Stories: How Quantifying, Classifying, and Formalizing Practices Shape Everyday Life*. Ithaca, NY: Cornell University Press.

Langston, Nancy. 2011. *Toxic Bodies*. New Haven, CT: Yale University Press.

Lash, Scott, Bronislaw Szerszynski, and Brian Wynne, eds. 1996. *Risk, Environment and Modernity: Towards a New Ecology*. Thousand Oaks, CA: Sage.

Latour, Bruno. 1986. "Visualization and Cognition: Thinking with Eyes and Hands." *Knowledge & Society* 6: 1–10.

Latour, Bruno. 1987. *Science in Action: How to Follow Scientists and Engineers through Society*. Milton Keynes, UK: Open University Press.

Latour, Bruno. 1988. *Pasteurization of France*. Translated by A. Sheridan and J. Law. Cambridge, MA: Harvard University Press.

Latour, Bruno. 2004. "How to Talk about the Body? The Normative Dimension of Science Studies." *Body and Society* 10 (2–3): 205–229.

Lindee, Susan M. 1994. *Suffering Made Real: American Science and the Survivors at Hiroshima*. Chicago: University of Chicago Press.

Lupandin, Vladimir M. 1998. "Chernobyl 1996: New Material concerning Acute Radiation Syndrome around Chernobyl." In *Research Activities about the Radiological Consequences of the Chernobyl NPS Accident and Social Activities to Assist the Sufferers by the Accident*, KURRI-KR-21, edited by Tetsuji Imanaka, 108–113. Kyoto, Japan: Kyoto University Research Reactor Institute.

Malko, Mikhail V. 1998a. "Assessment of the Chernobyl Radiological Consequences." In *Research Activities about the Radiological Consequences of the Chernobyl NPS Accident and Social Activities to Assist the Sufferers by the Accident*, KURRI-KR-21, edited by Tetsuji Imanaka, 65–89. Kyoto, Japan: Kyoto University Research Reactor Institute.

Malko, Mikhail V. 1998b. "Chernobyl Accident: The Crisis of the International Radiation Community." In *Research Activities about the Radiological Consequences of the Chernobyl NPS Accident and Social Activities to Assist the Sufferers by the Accident*, KURRI-KR-21, edited by Tetsuji Imanaka, 5–17. Kyoto, Japan: Kyoto University Research Reactor Institute.

Marples, David R. 1986. *Chernobyl and Nuclear Power in the USSR*. Basingstoke, UK: Macmillan.

Marples, David R. 1988. *The Social Impact of the Chernobyl Disaster*. Basingstoke, UK: Macmillan.

Marples, David R. 1996a. *Belarus: From Soviet Rule to Nuclear Catastrophe*. New York: St. Martin's Press.

Marples, David R. 1996b. "Chernobyl Disaster: Its Effect on Belarus and Ukraine." In *The Long Road to Recovery: Community Responses to Industrial Disaster*, edited by James K. Mitchell, 183–230. New York: United Nations University Press.

Marples, David R. 1999. *Belarus: A Denationalized Nation*. Amsterdam, The Netherlands: Harwood Academic.

Mathews, Andrew S. 2005. "Power/Knowledge, Power/Ignorance: Forest Fires and the State in Mexico." *Human Ecology* 33 (6): 795–820.

Matsko, V. P., and Tetsuji Imanaka. 1998. "Legislation and Research Activity in Belarus about the Radiological Consequences of the Chernobyl Accident: Historical Review and Present Situation." In *Research Activities about the Radiological Consequences of the Chernobyl NPS Accident and Social Activities to Assist the Sufferers by the Accident*, KURRI-KR-21, edited by Tetsuji Imanaka, 28–39. Kyoto, Japan: Kyoto University Research Reactor Institute.

Mazur, Allan. 1987. "Putting Radon on the Public's Risk Agenda." *Science, Technology, and Human Values* 12 (3, 4): 86–93.

McGarity, Thomas O., and Wendy Wagner. 2010. *Bending Science: How Special Interests Corrupt Public Health Research*. Cambridge, MA: Harvard University Press.

Medvedev, Zhores. 1979. *Nuclear Disaster in the Urals*. Translated by George Saunders. New York: W. W. Norton.

Medvedev, Zhores. 1990. *The Legacy of Chernobyl*. New York: W. W. Norton.

Michaels, David. 2008. *Doubt Is Their Product: How Industry's Assault on Science Threatens Your Health*. New York: Oxford University Press.

Miller, Robin. 1994. "School Students' Understanding of Key Ideas about Radioactivity and Ionizing Radiation." *Public Understanding of Science* 3: 53–70.

Mirny, S. 2000. "Information and Communication Factors' Impact on the Chernobyl Zone (1986–1990) Mitigation Workers' Health." ECOHSE Symposium, Kaunas, Lithuania, October 4–7.

Mirny, S. 2009. "*Chernobyl kak informacionnaya travma*" [Chernobyl as Information Trauma]. In *Travma: Punkty* [Trauma: Items], edited by S. Oushakine and Elena Trubina, 209–246. Moscow: Novoe Literaturnoe Obozrenie.

Mol, Annemarie. 2002. *The Body Multiple: Ontology in Medical Practice*. Durham, NC: Duke University Press.

Mould, R. F. 1988. *Chernobyl: The Real Story*. New York:: Pergamon Press.

Mould, R. F. 2000. *Chernobyl Record: The Definitive History of the Chernobyl Catastrophe*. Philadelphia, PA: Institute of Physics.

Murphy, Michelle. 2006. *Sick Building Syndrome and the Problem of Uncertainty: Environmental Politics, Technoscience, and Women Workers*. Durham, NC: Duke University Press.

National Academy of Sciences of Belarus (NASB). 1999. "Evgeni Fedorovich Konoplya. On the 60th Anniversary." *Proceedings of the National Academy of Sciences of Belarus* 1: 133–134. http://nasb.gov.by/rus/publications/vestib/vbi99_lb.php.

National Academy of Sciences of Belarus (NASB). 2010. "Evgenii Pavlovich Demid-chik: On the 85th Anniversary." *Proceedings of the National Academy of Sciences of Belarus* 1: 120–121. http://nasb.gov.by/rus/publications/vestim/vmd10_1a.php.

National Research Council Committee to Assess Health Risks from Exposure to Low Levels of Ionizing Radiation (NRC). 2006. *Health Risks from Exposure to Low Levels of Ionizing Radiation: BEIR VII, Phase 2*. Washington, DC: National Academies Press.

Nesterenko, Vassily B. 1998. *Chernobyl'skaya katastrofa: Radiacionnaya zashchita naseleniya* [Chernobyl Catastrophe: Radiation Protection of Population]. Minsk, Belarus: Belrad.

Nesterenko, Vassily B. 1999. *Recommendacii po radiacionnoy zashchite naseleniya i ih effektivnosti* [Recommendations on Radiation Protection of Population and Their Efficiency]. Minsk, Belarus: Belrad.

Nesterenko, Vassily B. 2001. "Preface." In *Radioactive Caesium and Intrauterine Fetus Development*, edited by Yuri I. Bandazhevsky, 3–6. Minsk: Belarus: Belrad.

Nesterenko, Vassily B. n.d. "Moi put'ot general'nogo kontstruktora mobil'noy atom-noy elektrostancii k radiacionnoy zashchite naseleniya i protivniku dal'neyshego primeneniya yadernyh tehnologii i ekspluatacii deistvuushchih AES" [My Path from Chief Engineer of a Mobile Nuclear Power Plant to Radiation Protection of Population and Opposition to Further Use of Nuclear Technologies and Operation of Existing Nuclear Power Plants]. In Vassily Nesterenko. *Sbornik statei i dokladov, 2001–2008* [Collection of Articles and Reports], 64–71. Minsk: Belrad.

Nesterenko, Vassily B., and Alexey Nesterenko. 2006. "Radio-Ecological Conse-quences in Belarus 20 Years after the Chernobyl Catastrophe and the Necessity of Long-Term Radiation Protection for the Population." In *Chernobyl: 20 Years On*, Report of the European Committee on Radiation Risk (ECRR), edited by Chris C. Busby and Alexey V. Yablokov, 185–226. Aberystwyth, UK: Green Audit Press.

Nesterenko, Vassily B., and Alexey Nesterenko. 2009. "Decorporation of Chernobyl Radionuclides." In *Chernobyl: Consequences of the Catastrophe for People and the Environment*, edited by Alexey V. Yablokov, Vassily B. Nesterenko, Alexey V. Nesterenko, and Janette D. Sherman, 303–310. Boston: Blackwell.

Nesterenko, Vassily B., Alexey V. Nesterenko, and Aleksandr Sudas. 2004. *Belarusian Experience in the Field of Radiation Protection of Population: Role of Governmental and Non-Governmental Structures in Solving These Problems*. http://www.ec-sage.net/D04-03.pdf.

Nesvetailov, Gennadi. 1995. "Changing Centre-Periphery Relations in the Former Soviet Republics: The Case of Belarus." *Social Studies of Science* 25 (4): 853–871.

Nikitchenko, Ivan. 1999. *Chernobyl: How It Happened*. Minsk, Belarus: Labor Rights Committee.

Norman, Donald A. 1998. *The Design of Everyday Things*. Cambridge, MA: MIT Press.

Nowotny, Helga, Peter Scott, and Michael Gibbons. 2001. *Re-Thinking Science: Knowledge and the Public in an Age of Uncertainty*. Cambridge, UK: Polity.

Oreskes, Naomi, and Erik M. Conway. 2010. *Merchants of Doubt: How a Handful of Scientists Obscured the Truth on Issues from Tobacco Smoke to Global Warming*. New York: Bloomsbury Press.

Ozasa, Kotaro, Yukiko Shimizu, Akihiko Suyama, Fumiyoshi Kasagi, Midori Soda, Eric J. Grant, Ritsu Sakata, Hiromi Sugiyama, and Kazunori Kodama. 2012. "Studies of the Mortality of Atomic Bomb Survivors, Report 14, 1950–2003: An Overview of Cancer and Noncancer Diseases." *Radiation Research* 177 (3): 229–243.

Paine, R. 1992. "Chernobyl Reaches Norway: The Accident, Science, and the Threat to Cultural Knowledge." *Public Understanding of Science* 1: 261–280.

Paine, R. 2002. "Danger and the No-Risk Thesis." In *Catastrophe and Culture. The Anthropology of Disaster*, edited by Susanna M. Hoffman and Anthony Oliver-Smith, 67–90. Santa Fe, NM: School of American Research Press.

Park, Chris. 1989. *Chernobyl: The Long Shadow*. New York: Routledge.

Peeples, Jennifer. 2011. "Toxic Sublime: Imaging Contaminated Landscapes" *Environmental Communication* 5(4): 373–392.

Petryna, Adriana. 2002. *Life Exposed: Biological Citizens after Chernobyl*. Princeton, NJ: Princeton University Press.

Petryna, Adriana. 2009. "Nuclear Payouts: Knowledge and Compensation in the Chernobyl Aftermath." *Anthropology Now* 1(2): 30–39.

Phillips, Sarah Drue. 2002. "Half-Lives and Healthy Bodies: Discourses on 'Contaminated' Food and Healing in Post-Chernobyl Ukraine." *Food and Foodways* 10: 27–53.

Powell, M., S. Dunwoody, R. Griffin, and K. Neuwirth. 2007. "Exploring Lay Uncertainty about an Environmental Health Risk." *Public Understanding of Science* 16: 323–343.

Proctor, Robert. 1995. *Cancer Wars. How Politics Shapes What We Know and Don't Know about Cancer*. New York: Basic Books.

Proctor, Robert, and Londa L. Schiebinger. 2008. *Agnotology: The Making and Unmaking of Ignorance*. Stanford, CA: Stanford University Press.

Rajan, Ravi S. 2002. "Missing Expertise, Categorical Politics, and Chronic Disasters. The Case of Bhopal." In *Catastrophe and Culture: The Anthropology of Disaster*, edited by S. M. Hoffman and A. Oliver-Smith, 238–259. Santa Fe, NM: School of American Research Press.

Read, Piers P. 1993. *Ablaze: The Story of the Heroes and Victims of Chernobyl*. New York: Random House.

Rotkiewicz, Marcin, Henryk Suchar, and Ryszard Kaminski. 2001. "Chernobyl: The Biggest Bluff of the 20th Century." *Wprost*, January 14.

Schmid, Sonja D. 2004. "Transformation Discourse: Nuclear Risk as a Strategic Tool in Late Soviet Politics of Expertise." *Science, Technology, and Human Values*. 29 (3): 353–376.

Schmidt, Kjeld and Carla Simone, 1996. "Coordination Mechanisms: Towards a Conceptual Foundation of CSCW Systems Design." *Computer Supported Cooperative Work* 5 (2/3): 155–200.

Scott, Joan W. 1992. "Experience." In *Feminists Theorize the Political*, edited by J. Butler and Joan W. Scott, 22–40. New York: Routledge.

Shcherbak, Iurii. 1989. *Chernobyl: A Documentary Story*. Translated by Ian Press. Basingstoke, UK: Macmillan.

Shevchouk, V. E., and V. L. Gourachevsky, eds. 2001. *15 Years after Chernobyl Disaster: Consequences in the Republic of Belarus and Their Overcoming*. National Report. Minsk, Belarus: State Committee on the Problems of the Consequences of the Catastrophe at the Chernobyl Nuclear Power Plant.

Shevchouk, V. E., and V. L. Gourachevsky, eds. 2003. *Chernobyl Consequences in Belarus*. National Report. Minsk, Belarus: Propilei.

Shevchouk, V. E., and V. L. Gourachevsky, eds. 2006. *20 Years after the Chernobyl Catastrophe: The Consequences in the Republic of Belarus and Their Overcoming*. National Report. Minsk, Belarus: State Committee on the Problems of the Consequences of the Catastrophe at the Chernobyl Nuclear Power Plant.

Skryabin, Anatolii M. 1997. "'Chelovecheski' faktor: Dozy i zashchitnye mery" [The "Human" Factor: Doses and Protective Measures]. *Ekologicheskaya Antropologiya*, annual edition: 47–51. Minsk: Belarussian Committee "Children of Chernobyl."

Soneryd, L. 2007. "Deliberations on the Unknown, the Unsensed, and the Unsayable? Public Protests and the Development of Third-Generation Mobile Phones in Sweden." *Science, Technology, and Human Values* 32 (3): 287–314.

Star, Susan Leigh. 1983. "Simplification in Scientific Work: An Example from Neuroscientific Research." *Social Studies of Science* 13 (2): 205–228.

Star, Susan Leigh. 1985. "Scientific Work and Uncertainty." *Social Studies of Science* 15 (3): 391–427.

Star, Susan Leigh. 1991. "Invisible Work and Silenced Dialogues in Knowledge Representation." In *Women, Work and Computerization*, edited by I. Eriksson, B. Kitchenham, and K. Tijdens, 81–92. Amsterdam: North Holland.

Star, Susan Leigh. 1995. "The Politics of Formal Representations: Wizards, Gurus, and Organizational Complexity." In *Ecologies of Knowledge. Work and Politics in Science and Technology*, 88–118. Albany: State University of New York Press.

Star, Susan Leigh. 2006. "Whose Infrastructure Is It, Anyway?" Unpublished paper.

Star, Susan Leigh, and Karen Ruhleder. 1996. "Steps toward an Ecology of Infrastructure: Design and Access for Large Information Spaces." *Information Systems Research* 7 (1): 111–135.

Star, Susan Leigh, and Anselm Strauss. 1999. "Layers of Silence, Arenas of Voice: The Ecology of Visible and Invisible Work." *Computer-Supported Cooperative Work* 8: 9–30.

Stepanov, Andrei. 2010. *Politika Chernobylya v Belarusi v 1986–2008 Godah: Formirovanie I Proyavleniya Diskurs-Koalicii.* [Politics of Chernobyl in Belarus, 1986–2008. Formation and Manifestations of Discourse-Coalitions]. PhD dissertation. European Humanities University.

Stephens, Sharon. 2002. "Bounding Uncertainty: The Post-Chernobyl Culture of Radiation Protection Experts." In *Catastrophe and Culture: The Anthropology of Disaster*, edited by S. M. Hoffman and A. Oliver-Smith, 91–112. Santa Fe, NM: School of American Research Press.

Strauss, Anselm L. 1970. "Discovering New Theory from Previous Theory." In *Human Nature and Collective Behavior: Papers in Honor of Herbert Blumer*, edited by T. Shibutani, 46-53. Englewood Cliffs, NJ: Prentice-Hall.

Strauss, Anselm L. 1985. "Work and the Division of Labor." *Sociological Quarterly* 26: 1–19.

Strauss, Anselm L. 1987. *Qualitative Analysis for Social Scientists.* Cambridge, UK: Cambridge University Press.

Strauss, Anselm L. 1988. "The Articulation of Project Work: An Organizational Process." *Sociological Quarterly* 29: 163–178.

Strauss, Anselm L., and Juliet Corbin. 1990. *Basics of Qualitative Research: Grounded Theory Procedures and Techniques.* Newbury Park, CA: Sage.

Strydom, Piet. 2002. *Risk, Environment and Society: Ongoing Debates, Current Issues and Future Prospects.* Philadelphia, PA: Open University Press.

Sullivan, Shannon. 2007. "White Ignorance and Colonial Oppression: Or, Why I Know So Little about Puerto Rico." In *Race and Epistemologies of Ignorance*, edited by Shannon Sullivan and Nancy Tuana. Albany: State University of New York Press.

Tchertkoff, Wladimir, dir. 2003. *Nuclear Controversies* [film].

Topçu, Sezin. 2013. "Chernobyl Empowerment? Exporting 'Participatory Governance' to Contaminated Territories" In *Toxicants, Health, and Regulation since 1945*, edited by Soraya Boudia and Nathalie Jas, 135–158. London: Pickering & Chatto.

Tulviste, Peeter., and James V. Wertsch. 1994. "Official and Unofficial Histories: The Case of Estonia." *Journal of Narrative and Life History* 4: 311–329.

United Nations Development Programme (UNDP) and United Nations Children's Fund (UNICEF). 2002. *The Human Consequences of the Chernobyl Nuclear Accident: A Strategy for Recovery.* A Report Commissioned by UNDP and UNICEF, with the Support of UN-OCHA and WHO, February 6. http://www.unicef.org/newsline/02chernobylstudy.htm.

United Nations Scientific Committee on the Effects of Atomic Radiation (UNSCEAR). 1988. *Sources, Effects and Risks of Ionizing Radiation.* Report to the General Assembly, with annexes. New York: United Nations.

United Nations Scientific Committee on the Effects of Atomic Radiation (UNSCEAR). 2000. *Sources and Effects of Ionizing Radiation.* Report to the General Assembly, with scientific annexes. New York: United Nations..

United Nations Scientific Committee on the Effects of Atomic Radiation (UNSCEAR). 2008. *Effects of Ionizing Radiation. Report to the General Assembly*, with scientific annexes, vol. 2. New York: United Nations.

Wallman, Sandra. 1998. "Ordinary Women and Shapes of Knowledge: Perspectives on the Contexts of STD and AIDS." *Public Understanding of Science* 7: 169–185.

Wertsch, James. 2002. *Voices of Collective Remembering.* Cambridge, UK: Cambridge University Press.

Williams, E. D., D. Becker, E. P. Demidchik, S. Nagataki, A. Pinchera, and N. D. Tronko. 1996. "Effects on the Thyroid in Population Exposed to Radiation as a Result of the Chernobyl Accident." In *One Decade after Chernobyl: Summing Up the Consequences of the Accident.* Proceedings of an International Conference, Vienna, April 8–12, edited by IAEA, 207–238. Vienna: International Atomic Energy Agency (IAEA), European Commission (EC), and World Health Organization (WHO).

Winner, Langdon. 1986. "On Not Hitting the Tar-Baby." In *The Whale and the Reactor: A Search for Limits in an Age of High Technology*, 138–154. Chicago: University of Chicago Press.

World Bank. 2002. *Belarus: Overview of the Consequences of the Accident on the Chernobyl NPP and Programs of Overcoming the Consequences.* Report No. 23883-BY. July 15.

World Health Organization (WHO). 1993. *Chernobyl: Pomogaet Vsemirnaya Organizaciya zdravoohraneniya* [Chernobyl: WHO Assistance]. Geneva: WHO.

World Health Organization, International Atomic Energy Agency, and United Nations Development Programme (WHO/IAEA/UNDP). 2005. "Chernobyl: The True Scale of the Accident: 20 Years Later a UN Report Provides Definitive Answers and

Ways to Repair Lives." Joint press release, September 5. http://www.who.int/
mediacentre/news/releases/2005/pr38/en.

Wynne, Brian. 1991. "Knowledge in Context." *Science, Technology, and Human Values* 16 (1): 111–121.

Wynne, Brian. 1992. "Misunderstood Misunderstanding: Social Identities and Public Uptake of Science." *Public Understanding of Science* 1: 281–304.

Wynne. Brian. 1993. "Public Uptake of Science: A Case for Institutional Reflexivity." *Public Understanding of Science* 2: 321–337.

Wynne, Brian. 1996. "May the Sheep Safely Graze? A Reflexive View of the Expert-Lay Knowledge Divide." In *Risk, Environment, and Modernity: Towards a New Ecology*, edited by Scott Lash, Bronislaw Szerszynski, and Brian Wynne, 44–83. Thousand Oaks, CA: Sage.

Wynne, Brian. 2003. "Seasick on the Third Wave? Subverting the Hegemony of Propositionalism: Response to Collins & Evans (2002)." *Social Studies of Science* 33 (3): 401–417.

Wynne, Brian. 2005. "Reflexing Complexity: Post-Genomic Knowledge and Reductionist Returns in Public Science." *Theory, Culture and Society* 22 (5): 67–94.

Wynne, Brian. 2008. "Elephants in the Rooms Where Publics Encounter 'Science'?: A Response to Darrin Durant, 'Accounting for Expertise: Wynne and the Autonomy of the Lay Public.'" *Public Understanding of Science* 17 (1): 21–33.

Yablokov, Alexey V. 2006. "The Chernobyl Catastrophe—20 Years After (a Meta-Review)" In *Chernobyl: 20 Years On. Health Effects of the Chernobyl Accident.* Report of the European Committee on Radiation Risk (ECRR), edited by Chris C. Busby and Alexey Yablokov, 5–48. Aberystwyth, UK: Green Audit Press.

Yablokov, Alexey V., Vassily B. Nesterenko, Alexey V. Nesterenko, and Janette D. Sherman (contributing editor). 2009. *Chernobyl: Consequences of the Catastrophe for People and the Environment.* Annals of the New York Academy of Sciences, 1181. Boston: Blackwell.

Yaroshinskaya, Alla. 1995. *Chernobyl: The Forbidden Truth.* Lincoln: University of Nebraska Press, 1995.

Yaroshinskaya, Alla. 1998a. "Impact of Radiation on the Population during the First Weeks and Months after the Chernobyl Accident and Health State of the Population 10 Years Later." In *Research Activities about the Radiological Consequences of the Chernobyl NPS Accident and Social Activities to Assist the Sufferers by the Accident*, KURRI-KR-21, edited by Tetsuji. Imanaka, 104–107. Kyoto, Japan: Kyoto University Research Reactor Institute.

Yaroshinskaya, Alla. 1998b. "Overview of Different Information about Acute Radiation Syndrome among Inhabitants around Chernobyl." In *Research Activities about the Radiological Consequences of the Chernobyl NPS Accident and Social Activities to Assist the Sufferers by the Accident*, KURRI-KR-21, edited by Tetsuji. Imanaka, 114–120. Kyoto, Japan: Kyoto University Research Reactor Institute.

Yaroshinskaya, Alla. 1998c. "Problems of Social Assistance to the Chernobyl Sufferers in Russia." In *Research Activities about the Radiological Consequences of the Chernobyl NPS Accident and Social Activities to Assist the Sufferers by the Accident*, edited by Tetsuji. Imanaka, 257–265. Kyoto, Japan: Kyoto University Research Reactor Institute.

Zaprudnik, Jan. 1993. *Belarus: At Crossroads in History*. Boulder, CO: Westview Press.

Zbarovski Z. I., T. F. Delin, M. V. Malko, I. I. Matveenko, V. B. Nesterenko, and Z. A. Rudak. 1995. "Zamechaniya po koncepcii zashchitnyh mer v vostanovitel'nyi period dlya naseleniya prozhivaushchego na territorii Respubliki Belarus'" [Comments on the concept of radiation protection during the rehabilitation period for populations residing in the Republic of Belarus]. December 21. Unpublished document.

Zgersky, M. 1998. "Legal Regime of the Chernobyl Problems in the USSR, Belarus, Russia, and Ukraine." In *Research Activities about the Radiological Consequences of the Chernobyl NPS Accident and Social Activities to Assist the Sufferers by the Accident*, KURRI-KR-21, edited by Tetsuji Imanaka, 266–270. Kyoto, Japan: Kyoto University Research Reactor Institute.

Index

234

Index

Printed in the United States
by Baker & Taylor Publisher Services

Printed in the United States
by Baker & Taylor Publisher Services